Traditional Foods and Nutritional Security

About the Author

Dr. Renu Agrawal is the Ex. Chief Scientist at CSIR-CFTRI, Mysuru, India and rural program coordinator for outreach activities at CFTRI. She has over 20 patents to her name, with 65+ publications in International and National Journals. She has presented more than 160 papers in various national and international conferences and delivered more than 650 talks. She has guided students for M.Sc, M. phil and PhD in Biotechnology and Microbiology. She has represented India as a team leader taking scientific delegation for the Asia meet at Indonesia, Bali nominated by DST, Government of India. She had her PhD from University of Rajasthan, Jaipur. Her areas of specialization include biotransformation and probiotics. Her research involves lactic acid bacteria, yeasts and *Bacillus* as probiotic properties inhibiting harmful diseases and reduce the use of antibiotics improving health in the society. Her group has developed greener probiotic technologies used by the common man. As CSIR-800 Rural development coordinator she led CSIR-CFTRI for CSIR 800 to give a better quality of life to 800 million people in India through scientific and technological intervention. Many awareness camps, demonstrations of technologies were conducted. She has been passionate about farmers and has conducted many national level conferences for all India farmers and for women farmers. She has also been proactive to show them alternate cropping with high nutrient crops. The microbial processes of her team on Ready to make idli and dosa batter are the most downloaded processes from CFTRI.

She has been awarded at multiple scientific associations. Some honourable mentions are Scientific fellow awards as FAMI,FNABS,FSAB,FISBT. She has been awarded Karamveer Chakra by REX-UN both bronze and silver, Best woman scientist NABS, Woman achiever from International Guild of woman achievers, Reflections Excellence Award from Rotary Innerwheel, Kissan Ratna puraskar, Abdul Kalam Lifetime achievement Award, Nari Shakti Award to name a few. She has been selected as the "Best woman scientist" by National Academy of biological sciences, felicitated by Rotary Inner wheel for scientific excellence with "Suguna Award" and felicitated at various scientific and societal forums. In Rotary Innerwheel she was awarded the Best secretary for two times and also Best ISO twice in the district 3180. She is the editor of six and reviewer for fifteen research Journals and has authored scientific books on Role of Probiotics in Improving Human Health, Utilization of Agricultural Wastes and Making Food Safe Free From Pathogens. The books are being co-published with CRC Press, UK. She has written many book chapters with Marcell Dekkar, USA and Taylor and Francis UK. Her work has won many best paper awards at the national and international conferences. She has taught as faculty of M.Sc Food Technology at CFTRI and as Professor to ACSIR students. She has served as the president of 'Association of Microbiologists of India", Mysore chapter. She has been on the board of many International Technical Advisory committees at Dubai, Bulgaria, Argentina and Indonesia. She has been on the National committees of NCERT of national Science Talent Search, selection of meritorious children under NCERT, scientist selection boards of UPSC, DST and DBT and many International technical committees. She is an Executive committee board member of "National Academy of Biological Sciences". She has been nominated as a National Experts Advisory Committee (NEAC) member from Government of India on; Scheme for Young Scientists and Technologists (SYST), Department of Science and Technology, Govt. of India, National children science congress, Ministry of science and Technology, member in TIASN (technological intervention for addressing societal needs (TIASN), ASAR, CODER, projects to be taken in cold and Desert areas, Pandit Deen Dayal Upadhyay Gram Sankal Pariyojna, AWSAR for augmenting writing skills, 'KIRAN" which involves projects for women. She has been nominated to the Consultative Group made by the Principal Scientific Advisor, Dr. Vijay Raghavan, cabinet rank, who reports to the prime minister of India for inputs on science and technological developments in the country. She has been appointed as a committee member of evaluators by DST on Smart village in Karnataka. She is a committee member with Karnataka government for Sustainable Development Goals. She has been nominated in the expert committee of Bangalore Chambers of Commerce and Industries. She regularly writes for City Today newspaper and Biostand Ups a science magazine. She has been honoured as "SHERO" by Biostandups, an online science magazine and is on their editorial board. She an advisor to the International journal IJBST, a publon publication. Dr. Renu has been leading innovations and change in organisations. Her programs and scientific interviews are telecasted on Doordarshan (DD science). Her views on women empowerment have been relayed on All India Radio on women's day both at the state and national radio. She is also involved in the upliftment of rural education of Karnataka through "Pratham" for teachers training workshops and in translating course material to be made into short videos for the children which will make science easy for them. She has been a family counsellor at Police helpline, Mysore for last twenty years and has saved many lives. She has been as a Jury member for the selection of an achiever for I- woman global awards under science and innovation.

Currently she is a science communicator, reviewer, author, visiting professor in many universities and colleges and a National Expert Advisor in many committees in the ministry of science and technology and training to KAS and KPS officers for Sustainable Development Goals, in Karnataka. She has edited the draft note writing of SDG goals on Food and Nutrition for Government of Karnataka and has written the nutrition policy for government of Karnataka to be under taken from 2021. Her commitment to work can be seen as she was taken as Project Advisor by CFTRI even after her retirement. She has been serving the society by her work. She has served as the secretary of "CSIR Pensioners Welfare Association, Mysuru centre after her retirement. She is a philanthropist and has been running many fellowships, national level competitions for poor meritorious students, novel writing competitions, constructed pre fixed auto stand in Mysore at Hardinge circle for public use and has donated a serum analyser and paediatrics kits for rapid detection of HIV to Asha Kirana hospital at Mysuru.

Traditional Foods and Nutritional Security

Renu Agrawal
Ex. Chief Scientist and Rural Program Coordinator
CSIR-Central Food Technology Research Institute
Mysuru, Karnataka
India

NEW INDIA PUBLISHING AGENCY
New Delhi-110 034

NEW INDIA PUBLISHING AGENCY
101, Vikas Surya Plaza, CU Block, LSC Market
Pitam Pura, New Delhi – 110 034, India
Email: info@nipabooks.com
Web: www.nipabooks.com

For customer assistance, please contact
Phone: + 91-11-27 34 17 17 Fax: + 91-11- 27 34 16 16
E-Mail: feedbacks@nipabooks.com

ISBN: 978-93-90175-94-9

Composed and Designed by NIPA.

National Academy of Biological Sciences [NABS]

NABS-Secretariat, Room No. 209, Second Floor, CAS in Botany, University of Madras
Guindy Campus, Chennai – 600 025, Tamil Nadu, India
E-mail: vapartha@gmail.com, secretarynabs@gmail.com / Mobile: 094471 73162
Visit: www.nabsindia.org

Dr. V. A. Parthasarathy, Ph.D., FNABS
President, NABS
Formerly: Director, ICAR-Indian Institute of Spices Research, Calicut
Emeritus Scientist, Indian Council of Agricultural Research
National Project Coordinator, Biodiversity International (CGIAR),TFT Project

Chennai – 600 096
05-05-2020

Foreword

Traditional foods have evolved endemically to various regions of the world depending on the local availability of food. These foods, depending on seasons also vary. Over centuries, people have been working on to improve its taste aroma and nutritional value. They have been also found to give immunity to the prevailing diseases in those regions. Even within a country like India, there are thousands of local foods being prepared in the kitchen. With globalisation and fast transport facilities, these foods have been carried to far off places and sold in restaurants as delicacies. Nutrition security plays an important role for good health and a strong immune system in order to fight the diseases. It would be very interesting to know the value of these traditional foods and their use in nutritional security.

It is heartening to note that a book entitled "Traditional Foods and Nutritional Security" has been authored by Dr. Renu Agarwal which aims at creating awareness in nutrition security management and the relation with traditional practices in food.

The book deals with the progress made on the ways of Nutrition security and different aspects of traditional foods. It is the various plant and animal systems which are the reservoirs of fibre, proteins, vitamins, minerals, anti- oxidants lipids, flavourants and essential phyto-chemicals. The intake of balanced diet can provide nutrition security. Consumers around the world have been preparing food items for balanced diet which has a direct link towards good human health.

Res: 32/482 C , Narmada Nilayam, Bharathan Bazar, Chelavoor, Calicut- 673 571, Kerala

The book has been divided into 13 chapters covering various aspects including those flavouring agents, the principle behind the preparations of the foods, functions of these foods, nutritional value, fortification of foods and policy issues.

This book provides a very comprehensive review starting from the introduction of nutrition security to the challenges by the industries. The book provides the most recent updates in the area with a holistic approach. The author has been working in the field of food science and microbiology since four decades. I am certain that the book will be useful to students, teachers and researchers interested in this topic.

V.A. Parthasarathy
Formerly Director
ICAR – Indian Institute of Spices Research
Kozhikode, Kerala 673 571

Preamble

The book aims at creating awareness in nutrition security management and the relation with traditional practices in food. It has also dealt with the progress made on the ways of Nutrition security and different aspects of traditional foods. It is the various plant and animal systems which are the reservoirs of fibre, proteins, vitamins, minerals, antioxidants, lipids, flavourants and essential phyto-chemicals. The intake of balanced diet can provide nutrition security. Consumers around the world have been preparing food items for balanced diet which has a direct link towards good human health.

Traditional foods have evolved endemically to various regions of the world depending on the local availability of food. These foods, depending on seasons, also vary. Over centuries, people have been working on to improve its taste, aroma and nutritional value. They have been also found to give immunity to the prevailing diseases in those regions. Even within a country like India, there are thousands of local foods being prepared in the kitchen. With globalisation and fast transport facilities, these foods have been carried to far off places and sold in restaurants as delicacies. Nutrition security plays an important role for good health and a strong immune system in order to fight the diseases. The book is very interesting in bridging the traditional foods and their use in nutritional security as it covers the coherence of food supply policies with respect to nutrition and food security. The book has been divided into 13 chapters covering various aspects including those flavouring agents, the principle behind the preparations of the foods, functions of these foods, nutritional value, fortification of foods and policy issues for Nutritional Security and the role of FAO and WHO. Importance of traditional Spices in Foods for Nutritional security has been written in a very comprehensive manner.

This book provides a very comprehensive review starting from the introduction of nutrition security to the challenges by the industries. The book will be useful to students, teachers and researchers interested in this topic.

Author

Contents

List of Abbreviations

ACTH	Adrenocorticotropic Hormone
ADA	American Dietetic Association
AI	Adequate Intake
AICR	American Institute for Cancer Research
AIV	African Indigenous Vegetables
ALA	Alpha Linolenic Acid
ATP	adenosine tri phosphate
BMI	Body Mass Index
BMZ	German Federal Ministry for Economic Cooperation and Development
CDC	Centers for Disease Control and Prevention
CFS	Committee on World Food Security
CGRP	Calcitonin gene related peptide
CHD	coronary heart disease
CIP	Clean-in-place
CO_2	Carbon dioxide
CSDH	Commission on Social Determinants of Health
CVD	Cardio vascular disease
DD	Diarrheal diseases
DEVCO	Directorate for International Cooperation and Development
DHA	Docosahexaenoic acid
DHS	Demographic and Health Survey
DNA	Deoxyribonucleic acid
DRI	Dietary Reference Intakes
EAR	Estimated average requirement
ECOSOC	Economic and Social Council
EFSA	European Food Safety Authority
EPA	Eicosapentaenoic acid
EU	European Union
FAO	Food and Agriculture Organization of the United Nations
FBO	Food business operators
FDA	Food and Drug Administration
FDI	Foreign Direct Investment
FFRC	Food Fortification Resource Centre
FIRST	Food and Nutrition Security Impact, Resilience, Sustainability and Transformation
FMM	Multipartner Programme Support Mechanism
FNB	Food and Nutrition Board
FNS	Food and Nutrition Security
FSN	Food security and nutrition
FSSAI	Food Safety and Standards Authority of India

g	Grams
GAFSP	Global Agriculture and Food Security Program
GEMs	Global Equipment Manufacturers
GI	Glycemic index
GIEWS	Global Information Early Warning System
GLUT	Glucose transporters
GTZ	German Agency for Technical Cooperation
GVA	Gross Value Added
H_2O	Water
HDL	High density lipoproteins
HGH	Human growth hormone
HPP	High Pressure Processing
HRAF	Human Right to Adequate Food
IBS	Irritable bowel syndrome
ICDS	Integrated Child Development Services
IDA	Iron-deficiency anemia
IFPRI	International food policy research institute
INR	Indian rupees
IOM	Institute for Medicine
kg	Kilograms
LA	Linoleic acid
LDL cholesterol	Low Density lipoproteins
MAFF	Ministry of Agriculture, Forestry and Fisheries
MDG	Millennium Development Goal
MDM	Mid-Day Meal
MoFPI	Ministry of Food Processing Industries
MUFA	Medium Chain Unsaturated Fatty Acids
NABARD	National Bank for Agriculture and Rural Development
NAIP	National Agriculture Investment Plans
NCD	Non-communicable diseases
NDA	Nutrition and Allergies
NFHS	National Family Health Survey
NGO	nongovernment organization
NHMRC	National Health and Medical Research Council
NIH	National Institutes of Health
O_2	Molecular oxygen
OECD	Organisation for Economic Cooperation and Development
OFSP	Orange-fleshed sweet potatoes
PDS	Public distribution system
PEA	Political economy analysis-
PMKSY	Pradhan Mantri Kisan Sampada Yojna
PROTA	Plant Resources of Tropical Africa
PUFA	Poly unsaturated fatty acids
PUFA	Polyunsaturated fatty acids-
RDA	Recommended Dietary Allowances
ROMA	RAPID Outcome Mapping Approach
SDG	Sustainable Development Goal
SME	Small and Medium Enterprises
SNP	Safety net programmes

SSFPE	Small-scale food processing enterprises
TIF	Traditional indigenous foods
UNECOSOC	United Nations Economic and Social Council
UNICEF	United Nations Children's Fund
USDA	US Department of Agriculture
UT	Union Territories
VAM	Vulnerability Analysis and Mapping
WCRF	World Cancer Research Fund
WFP	World Food Programme
WHO	World Health Organizations

1

Introduction

The human body consists of trillions of cells and to be healthy it needs variety of foods and nutrients. The food around the globe includes a variety of spices which have many health benefits. It has been found that spices and vegetables in the diet are very important for living healthy. The traditional foods, nuts, seeds, wild fruits and vegetables which are used in the diet are grown in specific areas. Traditional food system plays a significant role towards maintaining the well-being and the health of people in any specific region. However, it has been observed that due to the changing food patterns of human beings the health is being deteriorated. Food is a major factor which plays an important role in everyday life. In India, the staple foods include pearl millet (bājra), rice whole-wheat flour (atta) and many varieties of lentils. These include masoor (most often red lentils), tuer (pigeon peas), urad (black gram) and moong (mung beans). Indian food is popular among the people because it is healthy. Most Indians are conscious of their health and love to eat healthy food. Many regions around the world prepare foods that contain spices which have many health benefits.

2

Aspects of Traditional Foods and Nutritional Security

According to Indian Philosophy (Bhat, 2012) the physical body is called Annamaya Kosha.A body, when well nourished grows well by absorbing the energies from the food consumed. Fasting is another aspect of traditional food system. It is meant to purify the mind and the body. The traditional food concept has been changed in our society to a large extent since earlier times. The various aspects of traditional food system has been recorded in ancient scriptures, literature and in oral narratives of folk. In southern India especially in Karnataka many dietary varieties are available. These varieties have many uncommon vegetables which have health benefits. It has been found that the key traditional staple food and food groups like nuts, seeds, wild fruits and vegetables which are used in the diet are area specific.Traditional food system plays a significant role in maintaining the well-being and health of the people. Traditional systems have not been given much importance which has resulted in the use of fewer species, decreased dietary diversity due to household food insecurity and consequently poor health status. In order to bring the change the knowledge of the traditional food system is very vital. If awareness is created for traditional food systems it can contribute a lot towards making a healthy society and a strong nation. The traditional knowledge of food is considered to be the best for any particular geographical condition. Changing food patterns can damage the good health of the society. So, it is vital to know the importance of good food habits of our own tradition and balanced diet (Bhat, 2012).

FAO and UN (2015) have given international fund for agricultural development and world food programme.FAO of UN (2009) declared a draft of the world summit on food security. Highlights of agriculture census (2010-11) has been given. Lisa (2003) has written a book on exotic appetites for better nutrition.

3

Functions of Food

We all eat food because it provides the nutrients that are needed for the various functions of the body. Energy is required for growth and repair of tissues, to maintain body temperature and for physical activity. Energy comes from foods which are rich in carbohydrates, proteins and fat. This provides energy to the body from other nutrients and helps to keep the skin and nervous system healthy. They help towards the various biological activity in the human body. These nutrients are grouped into carbohydrates, proteins, lipids (fat), water, vitamins and minerals. These are used for growth, maintenance, repair and other needs of the body. Protective foods are which contain adequate amounts of vitamins, minerals and high-quality proteins which protect against the development of a deficiency disease. There are various chemical reactions taking place within each cell of a living organism. These provide energy for vital processes and for synthesizing new organic materials. This process is known as Metabolism. Living organisms get energy from the environment and use it for movement, growth, development and reproduction. The chemical reactions are enzyme-mediated. Hundreds of coordinated, multistep reactions, fueled by energy obtained from nutrients and/or solar energy, converts the readily available material into the molecules required for growth and maintenance. Meredith and Kaiser (2019) has described on how food can affect blood sugar.

At the cellular level the chemical processes in all living matter, animals, plants, fungi or bacteria are the same. All living matter are made up of large molecules called proteins, which provide support and coordinated movement. It also provides storage and transport of small molecules. There are also catalysts which enhances the chemical reactions. There are 20 essential amino acids, the combination of which forms the specific proteins. The protein molecules involved in performing similar functions in different organisms consist of the same sequences of amino acids. There are four nitrogen bases which synthesize DNA. These are adenine, guanine, cytosine and thymine. The different combinations of these bases synthesize the cell components. RNA (ribonucleic acid) contains the sugar ribose instead of the sugar deoxyribose and uracil as the nitrogen base instead of the base thymine. Green plants use the energy of sunlight to convert water (H_2O) and carbon dioxide (CO_2) to carbohydrates (sugars and starches), other organic (carbon-containing) compounds and molecular oxygen (O_2). Living organisms eat carbohydrates and other organic materials, by utilizing the oxygen synthesized by the plants to form water, carbon dioxide and energy. The process that removes hydrogen atoms (containing electrons) from the carbohydrates and carries them to the oxygen provides energy. In plants, animals, fungi and bacteria most of the steps in the conversion of carbon dioxide to carbohydrates are the same.

Fish oil from cold-water fish, such as salmon, mackerel, tuna, cod and herring have been found to contain large amounts of omega-3 fatty acids which helps to reduce inflammation by blocking inflammatory receptors. Among the traditional plants, consumption of Indian frank incense or *Boswellia serrata* has been found to have anti-inflammatory properties.

Consumption of tumeric and curcumin have shown positive effects in the treatment of arthritis and inflammation.However, with blood thinners, such as warfarin, turmeric should not be taken. Probiotics are present in yogurt and pickles, and they may help in the reduction of inflammation. A 2014 study found that administering *Lactobacillus casei* 01, a probiotic to the people with inflammation for 8 weeks resulted in the improvement. The most common foods are yogurts and pickles that are rich in probiotics. They are also available as supplements (Renu Agrawal, 2016). People with low immunity may suffer with side effects or interfere with other treatments.The U.S. Food and Drug Administration (FDA) do not regulate herbs and supplements and therefore it is important to be careful. Beckeman and Skjoldebrand (2007) looked into the networks to promote food innovations.

4

Nutritional Requirements for the Human Body: Working of Proteins, Fats, Carbohydrates, Minerals and Vitamins for Health, Nutrition and Balanced Diet.

We get energy from food. We need energy for all the activities which we get by eating. The body structures as muscles, organs and bones are composed of the nutrients contained in food which are absorbed by the body.

Nutritional requirement of the major nutrients

Nutrients are the substances which are found in food and drive biological activity. They are essential for the human body. These are categorized into proteins, fats, carbohydrates (sugars, dietary fiber), vitamins and minerals (Sugishima, 2019). They perform the vital functions like building all parts of the body as muscle, bone, teeth and blood. Out of the 20 different kinds of amino acids that make up proteins nine are not synthesized in the body and are therefore called essential amino acids. These essential amino acids have to be supplemented through food. High-protein, low-carbohydrate diets have been found to be effective.Protein is an important component of every cell in the body. Large part of hair and nails are made up of proteins. The human body uses protein to build and repair the tissues, helps in making enzymes hormones and other body chemicals. Protein is an important building block of bones, muscles, cartilage, skin and blood. Fats, carbohydrates and proteins are known as "macronutrient", as the body requires them in large amounts. Vitamins and minerals, are required in small quantities and are therefore called as micronutrients. Most of the times, the body breaks down the same amount of protein which is used to build and repair tissues. During illness, pregnancy and breastfeeding the amount of protein breaking down is high. Protein requirement increases after any injury or surgery, older adults and athletes.The body needs protein depending upon the health and activity level of a body.All enzymes are proteins and they help in the biochemical reactions in and outside of the body cells.The structure of enzymes allows them to combine with other molecules inside the cell called substrates. This helps to catalyze the reactions which are essential for cell metabolism.Digestive enzymes like lactase and sucrase function outside the cell and help to digest sugars. Some enzymes require molecules like vitamins / minerals for a reaction in a cell. Enzymes are required for the digestion, energy production, blood clotting and muscle contraction or it can cause a disease. =

Metabolism in the body with Nutrients

Charlotte and Dode (2016) have discussed on the nutrition and food systems. US dietary guidelines (2015-2020) have been discussed well.

Requirement and metabolism of proteins

Some proteins are hormones which help in the communication between the cells, tissues and organs.These are made and secreted by the endocrine glands and are then transported by the blood to various organs where they bind to the protein receptors on the cell surface.Hormones are of three types: Protein and peptides are made from chains of amino acids, ranging from a few to several hundred. Steroids are made from the fat cholesterol. The sex hormones, testosterone and estrogen, are steroid-based. Amines are made from the individual amino acids tryptophan or tyrosine, which help in the making of hormones related to sleep and metabolism. Protein and polypeptides make up most of the body's hormones.

Glucose or sugar is taken up by insulin into the cell.Glucagon helps in the breakdown of stored glucose in the liver. hGH (human growth hormone) increases the growth of many tissues and bones.Reabsorption of water in the kidneys is done by ADH which is an antidiuretic hormone. Release of cortisol, is done by ACTH (adrenocorticotropic hormone). The fibrous protein gives the cells and tissues some stiffness and rigidity.These proteins include keratin, collagen and elastin, which helps in forming the connective framework of many structures in the body.Keratin is a structural protein that is found in the skin, hair and nails. Collagen is the most abundant protein in the body.It is the structural protein of bones, tendons, ligaments and skin. Elastin is many times more flexible than the collagen. Its high elasticity allows many tissues in the body to return to their original shape after stretching or contracting which includes; uterus, lungs and arteries. Fibrous proteins provide with structure, strength and elasticity. It is the proteins that play a major role to regulate the levels of acids and bases in the blood and other body fluids. The balance between acids and bases is measured using the pH scale. The pH 2 is of stomach acid. If the pH of some major food items is looked into it is found that the pH of tomato juice is 4; black coffee, pH 5; human blood, pH 7.4 and milk of magnesia, pH

10. Eating food of different pH values could be harmful (Aranceta et al. 2011). However, it is the proteins which regulate the body pH. Hemoglobin binds with acid to maintain the normal pH of the blood. Phosphate and bicarbonate are the other buffer systems (Agence Française de Sécurité Sanitaire des Aliments (AFSSA, 2010). Albumin and globulin are blood proteins that help to maintain the body's fluid balance by attracting and retaining water (National Health and Medical Research Council, Department of Health and Ageing (2013), Food and Agriculture Organization of the United Nations (FAO) &World Health Organization (WHO) (2010). When the proteins are unbalanced in the body the levels of albumin and globulin decreases. These low levels are not able to keep the blood in the blood vessels. The fluid is thus forced into the spaces between the cells. As the fluid continues to build up in the spaces between the cells, swelling or edema occurs, particularly in the stomach region (Siri-Tarino et al., 2010). Protein malnutrition is known as kwashiorkor when a person consumes enough calories but not enough proteins. Proteins also increases the immune health.Proteins are helpful in forming immunoglobulins and antibodies that Fight infections (European Commission, DG Joint Research Centre (2013), Elmadfa (2009).Antibodies are proteins in the blood that protect the body from harmful bacteria and viruses (Harika et al., 2013, Micha et al., 2014).As a result, the body develops immunity against the diseases. Transport proteins, carry substances through the bloodstream in and out of cells or within the cells. The substances transported by these proteins include nutrients like vitamins or minerals, blood sugar, cholesterol and oxygen (EUFIC website, 2014, Ebbert and Jensen (2013); Thomas et al., 2012). Glucose transporters (GLUT) move glucose to the cells, while lipoproteins transport cholesterol and other fats into the blood.Protein transporters are specific, as they bind only to the specific substances (Nedergaard et al, 2007; Lidell et al., 2013).Proteins also have storage roles namely, ferritin is a storage protein that stores iron (European Commission, Joint Research Centre, 2011). Another storage protein is casein, the principal protein in milk that helps babies to grow. Some proteins transport nutrients to the entire body, while others store them.Proteins can also supply energy to the body. One gram of protein contains four calories, the same amount of energy that carbohydrates provide. Fats supply the maximum energy which is nine calories per gram (World Health Organization (WHO) Media Center (2013). In fact, under normal circumstances, the energy needs supplied to the body by proteins is very little.However, in a state of fasting (18–48 hours of no food intake), the body breaks down skeletal muscles for the amino acids to supply energy (Jakobsen et al., 2009; Mensink et al, 2003). Muscles, skin and bones in the human body contain significant amounts of proteins which includes enzymes, hormones, antibodies and also work as neurotransmitters. Hemoglobin, which is a carrier of oxygen in the blood, is a protein and essential

for the functioning of every cell in the body. The structures differ according to the sequence in which the amino acids combine.The 20 different amino acids that the body uses to synthesize proteins include; alanine, arginine, asparagine, aspartic acid, cysteine, glutamic acid, glutamine, glycine, histidine, isoleucine, leucine, lysine, methionine, phenylalanine, proline, serine, threonine, tryptophan, tyrosine and valine.Amino acids are organic molecules that consists of carbon, hydrogen, oxygen, nitrogen and sulfur. Plant foods like peas, beans and grains mostly contain incomplete protein. Complementary proteins are present in foods which contain incomplete proteins that people can combine to supply complete protein. Examples include rice and beans or bread with peanut butter.The nine essential aminoacids that the human body does not synthesize are histidine, isoleucine, leucine, lysine, methionine, phenylalanine, threonine, tryptophan and valine. Foods that contain these nine essential acids are termed to have complete proteins. Animal sources for complete proteins are present in milk, meat and eggs.Soy and quinoa are vegetable sources of complete protein. Complete protein can be obtained if red beans or lentils are combined with rice or wholemeal bread. Intake of enough proteins helps in the metabolism. According to the Dietary Guidelines for Americans (2015–2020) the recommendation is between 10 and 35 percent of an adult's daily calories must be obtained from proteins. For children, it should be 10 to 30 percent.According to the Centers for Disease Control and Prevention (CDC), men in the United States get 16.1 percent of their calories from protein on an average and women 15.6 percent. Lack of proteins in the diet can lead to problems of malnutrition, such as kwashiorkor and marasmus which can be fatal.Even eating disorders like anorexia nervosa occurs due to the deficiency of proteins. Diseases like irritable bowel syndrome (IBS) or gastric bypass surgery are caused due to very low protein intake.This makes the muscle tone weak, edema, thin and brittle hair, skin lesions, loss of muscle mass in children and even stunted growth. Biochemical tests have shown low serum albumin and hormonal imbalances.

It is not necessary to have proteins from meat alone. Seafood, eggs, pulses and beans provide sufficient proteins.According to FDA an adult requires 50 grams of protein per day. So a 2, 000-calorie diet should be sufficient. The requirement per day, changes with age, gender, activity level and pregnancy.The proportion of amino acids available in specific protein foods and their digestibility is an important factor.It also remains unclear how protein metabolism affects the need for protein intake.According to the U.S Department of Agriculture (USDA), One ounce lean meat, poultry, seafood, one egg, one tablespoon of peanut butter, half an ounce of nuts, seeds, one fourth of a cup of cooked beans or peas provide around 1 ounce of protein / serving.Accordingly, consuming 5 and 7 ounces of proteins in a day for most people over the age of 9 years is sufficient. One gram of protein contains 4 calories whereas one gram of fat has 9 calories.An average person in USA consumes around 16 percent of their calories from protein. Supplementation of high fiber foods such as fruits, vegetables and whole grains with protein foods could have a negative effect. Athletes use protein shakes and supplements. In recent times, a wide range of protein supplements are available. Too much protein intake could increase the risk of osteoporosis or kidney problems.In a study it has been seen that whey proteins could affect glucose metabolism and muscle protein synthesis. Other research concludes that at least one type of whey supplement can reduce body fat and preserve muscles with a reduced-calorie diet. As supplements, whey protein and similar products do not have FDA approval. Cooking methods like grilling do not add extra fat, wholemeal rather than white bread and unsweetened peanut butter are good, plant-based proteins like beans, lentils and soy products are good to eat. If protein levels become very high it can damage a person's health, including kidney and may cause dehydration (Yvette, 2018).

It has been found that a vegan diet rich in fresh fruits and vegetables, whole grains, legumes, nuts and seeds significantly reduced systemic inflammation. These are the most versatile molecules for the human body and are key to almost all biological processes. The average RDA (recommended dietary allowance) for protein is calculated using the ratio of 1 gram of protein for every 1 kilogram of a person's body weight. The recommended dietary allowance or RDA for protein varies depending upon the age, gender, pregnancy and breast feeding, activity levels. Adults must eat 0.8 g per kilograms (kg) of body weight daily. Among various nutrients, protein, lysine, tryptophan, iron, zinc, vitamin A and vitamin C are essential for human nutrition and their deficiency leads to various symptoms and health disorders. Kraemer and Zimmermann (2007) have studied nutritional anemia. Erucic acid, glucosinolates and Kunitz trypsin inhibitor (KTI) are the antinutritional factors. Their consumption in higher levels leads to adverse effects in humans and livestock. Humans require 0.66 g protein/kg body weight/day to meet the requirement for proper growth and development (WHO; 2007. Joint WHO/FAO/UNU Expert Consultation). The daily lysine requirement is 30 mg/kg body weight/day for adults, while it is 35 mg/kg body weight/day for children of 3-10 yr of age (WHO; 2007. Joint WHO/FAO/UNU Expert Consultation). Deficiency leads to fatigue, dizziness, nausea, anaemia, delayed growth, loss of appetite and reproductive tissue (Tome and Bos, 2007; Moehn et al., 2012). Tryptophan is a building block of proteins and functions as precursors for several neurotransmitters and regulators of metabolic pathways.Tryptophan is required at the rate of 4 mg/kg body weight/day in adults and 4.8 mg/kg body weight/day in children (3-10 yr; WHO; 2007. Joint WHO/FAO/UNU Expert Consultation). The deficiency leads to depression, anxiety and impatience. Weight loss and slow growth in children are the major symptoms of tryptophan deficiency (Tome and Bos, 2007; Moehn et al., 2012).

According to the Institute for Medicine (IOM), the daily RDA for protein are given in Table(1).

Table 1: The daily RDA for Protein.

Life stage and gender	RDA in grams (g) per day
Infants and children	
0–6 months	9.1
6–12 months	11.0
1–3 years	13.0
4–8 years	19.0
Males	
9–13 years	34.0
14–18 years	52.0
19–70 years and older	56.0
Females	
9–13 years	34.0
14–70 years and older	46.0
Pregnant or breastfeeding women	
All ages	71.0

Source: Institute for Medicine (IOM), Dietary reference intakes. Nat. Acad. Press, Washington, USA.

Being physically active can increase the RDA of protein amount to be eaten by the people. A 2016 study recommends eating: 1.0 g of protein per kg of body weight with minimal activity levels, 1.3 g of protein per kg of body weight with moderate activity levels, 1.6 g of protein per kg of body weight with intense activity levels, Anyone who is pregnant or breastfeeding will need to eat a lot more protein than other people.Some studies have also found that people may need to increase the protein intake as they age.A 2016 study concluded that older adults should eat more protein than is currently recommended to promote healthy aging. The researchers recommended that adults should ideally consume protein in the range of 1.2–1.6 g per kg of body weight daily, to prevent age-related muscle loss or sarcopenia.The study also concluded that these amounts would also improve appetite control, satiety and weight management.Side effects of too much protein can cause intestinal discomfort and indigestion, dehydration, unexplained exhaustion, nausea, irritability, headache and diarrhea. People can typically consume 2g of protein per kg of their body weight daily for a long-term, without any significant side effects. Some people, such as elite athletes, may be able to eat as much as 3.5 g per kg of body weight daily without any side effects.Most research indicates that eating more than 2 g per kg of body weight daily of protein for a long time can cause health problems. There are serious risks associated with chronic protein overconsumption, including:cardiovascular disease, blood vessel disorders, liver and kidney injuries, seizures and death.Doctors have also linked certain conditions to chronic protein overconsumption: type 2 diabetes, cancer, osteoporosis and osteopenia. The IOM recommends people get between 10 and 35 percent of their daily energy intake from protein. Researchers have associated higher protein diets with weight loss.High-protein diets promote weight loss as they give a feeling of fullness. This reduces hunger cravings and overeating. In a study done on adolescent overweight or obese girls it was found that eating breakfast high in protein, might control neural signals that regulates food cravings.However, more research is needed to understand the full relationship between high protein diets and weight loss. In order to maintain a healthy diet all protein-rich foods are not good. Best protein-rich, low-calorie foods include; skinless chicken breast 3 ounces (oz) (less than 26 g protein and 113 calories), 1 scoop of whey protein (less than 24–26 g and 130 calories), 6 oz greek yogurt (less than 17 g protein and 100 calories), 2 large eggs (less than 12 g protein and 144 calories), tofu ½ cup (less than 10 g protein and 95 calories), 2 tablespoons (tbsp) peanut butter (less than 8 g protein and 190 calories), cup ½ beans (less than 8 g protein and 110 calories), almonds 1 oz (less than 6 g protein and 165 calories), cooked oatmeal 1 cup (less than 6 g protein and 165 calories). Quinoa is found to be very good. Cooked ½ cup quinoa gives less than 4 g protein and 110 calories.Side effects from protein overconsumption may contribute to kidney and liver conditions.Some people cannot eat as much protein as others because of conditions that interfere with digestion. According to national statistics gathered between 2011 and 2014, most American adults of 20 years of age or more only consumed between 15.6 and 16.1 percent of their daily energy intake from protein.Eating a protein-rich diet can help in losing weight as it avoids overeating. It builds lean muscle which helps to burn more calories throughout the day. The best high protein foods if someone wants to lose weight are:

1. Black beans are inexpensive Source of protein. Black beans can be prepared in a variety of ways, making them a very versatile ingredient when preparing meals.
2. Lima beans provide about 21 grams (g) of protein per 100 g serving.
3. Yellow corn has about 15.6g of protein per cup. Additionally, corn also contains a good amount of fiber and mineral like calcium.
4. Salmon contains omega-3 fatty acids and is an good Source of protein.
5. Potatoes are good Sources of nutrients, including protein. One medium potato with the skin contains more than 4g of protein.

6. One cup of raw broccoli contains almost 2.6 g of protein and has a variety of nutrients like folate and potassium.
7. Cauliflower has a lot of protein and very less calories. One cup of chopped cauliflower has 27 calories and 2 g of protein.
8. Chinese cabbage / bok choy, has good amount of proteins and is full of antioxidants.
9. Eggs are an excellent source of protein, nutrients and healthful fats.
10. Beef provides high amounts of protein per serving. Persons suggested with a moderate carbohydrate diet can eat lean beef and people on a low-carb diet can eat fattier beef.
11. Chicken breast is a lean source of protein. A 136 g of skinless chicken breast gives around 26 g of protein.
12. Oats provide 17g of protein per 100g. They also provide carbohydrates. Raw oats are easy to prepare as oatmeal and can be supplemented with fruits and nuts.
13. Tuna is an excellent and widely available Source of protein and has a low calorie count. It is a lean fish with minimal fat. It can be added to salads, sandwiches and snacks. Dressings, like mayonnaise can add unwanted calories.
14. Tempeh is a popular source of protein for vegetarians and vegans providing about 17g per half cup.Tempeh comes from soybeans, like tofu.
15. Spirulina is a blue green algae that grows in both fresh and salt waters. It offers a variety of nutrients and protein.
16. Legumes are both high in fiber and protein and helpful in weight loss. Some people find difficulty in digesting legumes.
17. Hemp seeds in salads offer about 9.5 g of protein per tablespoon.
18. Sun-dried tomatoes are a good source of protein, as well as additional nutrients and fiber.
19. Guava is a tropical fruit and is one of the most protein-rich fruits available. It also provides vitamin C.
20. Artichokes are high in fiber and offer a good amount of protein.
21. Peas are high in protein, fiber and other nutrients. Peas are inexpensive, and are used in lots of recipes.

22. Bison meat is another excellent source of protein. Bison is lean meat, offering less fat per serving than beef.
23. Lean pork is a good source of protein. Pork roasts and tenderloin are good choices for meals.
24. Turkey packs a powerful punch of protein. Boneless turkey can provide about 13 g of protein per 100 g.
25. Chickpeas are a healthful vegetarian protein that is high in fiber, and full of nutrients that support heart, has anti cancer properties and enhances bone health.
26. Quinoa is one of the only complete sources of vegetarian protein. Quinoa contains all 11 essential amino acids needed and is an excellent choice for vegetarians and vegans.
27. Plain, low-fat Greek yogurt provides 19 g of protein / 200g. Plain yogurt is better than with some fruits or seeds.
28. Cottage cheese has a high amount of protein, calcium and other nutrients.
29. Almonds eaten in dry roasted form or raw is a good filling and protein-rich snack.
30. Cow's milk is an excellent Source of protein for people that can tolerate drinking milk. An 8 ounce serving of milk contains 8 g of protein.
31. Lentils pack a hefty dose of plant protein and fiber. They are economical and promotes heart health.
32. Pumpkin seeds are full of protein and minerals, like magnesium and selenium. Dry roasted seeds are good for health.
33. Avocados contain healthy fats, as well as protein.Avocados contains protein, unsaturated fat, fiber and nutrients, like potassium. It should be eaten in small quantities due to its high calorie level.
34. Pistachios are a low calorie nut and contains a big serving of protein. One ounce of pistachios contains about 6g of protein along with other nutrients and high level of B-6.
35. Chia seeds contains more than 5gof proteins per ounce, along with omega-3 fatty acids, fiber and calcium. It can be added to smoothies / salads for health benefits.
36. Nut butters, like peanut butter, contain a lot of calories, but a portion-controlled serving can add unsaturated fat and proteins.

37. Halibut is a white fish an excellent Source of lean protein which contains 30g of protein in half a fillet.
38. Asparagus is a rich source of protein. It is also full of nutrients, including B vitamins and is low in carbohydrates.
39. Watercress is a cruciferous vegetable which grows in water. It has high protein content and contains vitamin K. Adding some watercress to salads can maximizes the health benefits.
40. Brussel sprouts are full of protein, fiber and vitamins. One cup contains 3 g of protein.
41. Spelt is a type of hulled wheat that has a very high protein content.
42. Teff is a grass which is ground to make flour. This gluten-free food has a fairly high protein content with about 13 g of protein per 100 g serving.
43. Whey protein powder is used by many bodybuilders and athletes as a supplement to increase muscle mass and strength. This powder is made from the proteins found in the liquid part of milk and provides a substantial amount of protein to a person's diet.Many protein Sources offer additional nutrients that benefit overall health as well.

Requirement and metabolism of Fats: Dietary fats are essential for the body energy and for supporting the cell growth. They protect the organs, absorb some nutrients, produce important hormones and help in keeping the body warm. Dietary fats are good for health and for optimal functioning of the human body. Dietary fats are a source of energy that functions as structural building blocks, carry fat-soluble vitamins, are involved in vital physiological processes in the body and are also utilized in a number of important biological functions like growth and development. Fat is the most concentrated source providing 9 kcal per 1 gram consumed, which is more than double the energy content of protein or carbohydrate (4 kcal per gram) and more than quadruple the energy content of fibre (2 kcal per gram). Fat can be stored in the body's fat tissue, which releases fatty acids when energy is required. The cell membrane separates the inside from the outside of the cell and controls the movement of substances in and out of the cells. They are mainly made up of phospholipids, triglycerides and cholesterol. Both length and saturation of the fatty acids from phospholipids and triglycerides affects the arrangement of the membrane and thus the fluidity. Shorter chain fatty acids and unsaturated fatty acids are less stiff and less viscous which makes the membranes very flexible. Important biological functions like the process of endocytosis get influenced. The brain is very rich in fat (60%) with a fatty acid composition containing docosahexaenoic acid (DHA). This is the major brain fatty acid. According to the European Food Safety Authority (EFSA, 2010a, b) the lipids of the retina also contain very high concentrations of DHA. Fat is a carrier for the fat-soluble vitamins A, D, E and K, and supports their absorption in the intestine. Consuming sufficient amounts of fatty foods that contain these vitamins is thus essential for adequate intake of these micronutrients. Our bodies cannot produce the polyunsaturated fatty acids (PUFA), linoleic acid (LA) and alpha linolenic acid (ALA). Without these essential fatty acids some important body functions may not take place. Therefore, it is important to consume them through the diet. LA and ALA can be converted to longer chain fatty acids and into compounds with hormone-like property (prostaglandins) or inflammatory properties (leukotrienes). As such, essential fatty acids are involved in many physiological processes such as blood clotting, wound healing and inflammation. Although the body is able to convert LA and ALA into the long chain versions arachidonic acid (AA), eicosapentaenoic acid (EPA), and, to a lesser extent, to docosahexaenoic acid (DHA), this conversion seems limited (Kaur et al., 2014). The longer chain fatty acids like EPA and DHA are known as conditionally essential and it is recommended to consume direct sources of these particular long chain fatty acids. The richest source of EPA and DHA is oily fish, including anchovy, salmon, tuna and mackerel. Mozaffarian et al., (2013) has studied on the mortality in older adults due to lipids.

World Health Organization (WHO) and the European Food Safety Authority (EFSA) have given recommendations differently as the results cannot be the same always as there may be diabetic women over the age of 65, the differences in the study duration, the dose and composition of the intervention. For dietary fats, the change in the type of fats consumed (saturated fat replaced by unsaturated fat) or changing the type in combination provides protection against cardiovascular diseases. Most authorities recommended intake for total dietary fat in adults as 20-35% of the total daily energy.

According to the WHO, CVD is the number one cause of death globally, accounting for 30% of total mortality (World Health Organization, WHO) (2013). In the 1970's, a link between total/saturated fat consumption and the risk for heart disease mortality was established in a study conducted in seven countries. After this, several authoritative bodies made recommendations to reduce saturated and total fat from the diet, to prevent CVD (Aranceta & Perez (2012). Recently, the studies have proved that a replacement of SFA by PUFA reduces the risk for CVD (Aranceta & Perez, 2012 and Siri-Tarino et al, (2010). Intervention studies on the effects of fat intake on CVD have mainly studied the effects of a reduction in total/saturated dietary fat (replaced by other nutrients) on the levels of blood lipids. If the level of blood lipids are abnormal it increases the risk factor for developing CVD (Lordon et al., 2018). High levels of LDL cholesterol gives a higher risk. Similar effects are observed with an increased ratio between LDL (or total) cholesterol and HDL cholesterol. This ratio is a good marker for CVD risk than LDL alone (Mensink et al.,2003). In addition, recent research indicates that Apo A1 and Apo B (proteins that are involved in lipid transport in the body) blood levels and the size of the LDL particle can be good risk markers for CVD (Barona and Fernandez, 2012). A smaller LDL particle size is more likely to induce atherosclerosis in which the plaque formation takes place on the inside of the blood vessels. This increases the risk of blockage and subsequent cardio vascular diseases take place.An elevated level of blood triglycerides is also linked to a higher risk of CVD (Jakobsen et al., 2009). In some intervention studies, replacement of saturated fat by digestible carbohydrates has been linked to a more atherogenic blood profile and dyslipidaemia (Siri-Tarino et al., 2010). If SFA are replaced with MUFA then, positive effects are observed on the blood pressure and blood lipid profile to a large extent (Siri-Tarino et al., 2010; Astrup et al., 2011). The effects of SFA on the blood lipid profile depends on the individual SFA, due to the different chain lengths of the fatty acids.More evidence is required to link any specific saturated fatty acid with a strong adverse effect on blood lipids or a disease endpoint (Fattore and Fanelli, 2013). SFA with a shorter carbon chain (lauric acid) do increase LDL cholesterol stronger than the ones with a longer chain(stearic acid) but at the same time the former has a higher HDL raising potential (Mensink et al, 2003 and Fattore and Fanelli, 2013) . The short-chain fatty acids may show more positive effects on the total/HDL cholesterol ratio (Mensink et al., 2003). A recent review shows that palmitic acid, the most abundant saturated fatty acid in the diet, seems to increase LDL cholesterol but also HDL cholesterol, and has not been shown to increase the risk for CVD (Fattore and Fanelli, 2013). With data on the relation between specific SFA and CVD endpoints lacking, the total body of evidence is insufficient to favour one SFA over another regarding CVD benefits and further research is needed to confirm any differences in health effects between these fatty acids (Fattore and Fanelli, 2013).

1. The adverse health effects of TFA, can be seen on the blood lipid levels and CVD (Brouwer et al., 2013; Mozaffarian et al., 2009 and Bendsen et al., 2011).They also initiate low-grade inflammation and predisposed to insulin resistance, decrease insulin sensitivity which is related to the development of type 2 diabetes (Kalupahana et al., 2011). A Cochrane review found a reduction of CVD risk in studies of fat modification.It was found that SFA when replaced by MUFA/PUFA) in men, (Hooper et al., 2012) replacing 5% of energy intake of saturated fat by PUFA, results in a 10% CVD risk reduction (Micha & Mozaffarian, 2010). Populations consuming a Western-type diet (high in refined grains, fat and sugar and

low in wholegrain) was found harmful (Drewnowski and Popkin, 1997). According to Astrup et al., (2011) the replacement of 1% of energy from saturated fat with PUFA lowers LDL cholesterol and is likely to produce a reduction in CHD incidence from 2 to 3%. The European Food Safety Authority (EFSA) Panel on Dietetic Products, Nutrition and Allergies (NDA) have concluded that there is a direct connection between the consumption of mixtures of dietary SFA and an increase in blood LDL-cholesterol concentrations, and that replacement of a mixture of SFA with cis-MUFA and/or cis-PUFA in foods or diets on a gram per gram basis reduces LDL cholesterol concentrations (European Food Safety Authority (EFSA, 2011). Persons with higher n-3 fatty acid levels lived around 2.22 years longer, after the age of 65 years (Mozaffarian et al., 2009). A study has shown the effects of Omega-3 supplements and cardiovascular diseases(Mohebi-Nejad and Bikdeli 2014) and coronary heart disease(Farvid etal., 2014) and it was found that linoleic acid, the main n-6 fatty acid, lowered the risk for both these endpoints.

Changing the types of fat (PUFA instead of SFA), rather than reducing the total amount of fat in the diet, may have a positive effect on glucose metabolism (Hu et al., 2001). Animal studies have shown improvement of several metabolic factors, including insulin sensitivity, underlying the development of type 2 diabetes when SFA are replaced with PUFA. Insulin sensitivity refers to the capacity of body cells to respond to the hormone insulin, which supports the uptake of glucose, amino acids and fatty acids. A 12-week n-3 PUFA supplementation in people with obesity, insulin resistant children and adolescents, showed positive effects on blood lipids and insulin sensitivity (Juárez-López et al., 2013). Recently, meta-analyses has not proved that the consumption of fish and n-3 fatty acid could lower the risk for type 2 diabetes in humans(Wu et al., 2012; Zhou et al., 2012). Insulin resistance and body response to fat intake has some relation (Lefevre et al., 2005) and is associated with an increased risk for CVD (Zhou et al., 2012). Inflammation in fat tissue of individuals is associated with the pathogenesis of insulin resistance and metabolic syndrome (Rosgvist et al., 2014). It could be due to the activation of innate immune processes by SFA. However, the n-3 fatty acids, EPA and DHA, have anti-inflammatory properties and can modulate adipose tissue inflammation. Reduction in adipose tissue helps in lowering the inflammation, improving fatty acid metabolism and insulin sensitivity.Particular diet intake helps in such diseases.

Obesity also increases the risk of different types of cancer. According to the World Cancer Research Fund and American Institute for Cancer Research evidence is lacking on this (World Cancer Research Fund (WCRF) / American

Institute for Cancer Research (World Cancer Research Fund (WCRF) / American Institute for Cancer Research (AICR, 2007)). The recent researches suggests that a higher level of n-3 fatty acid consumption may be associated with reduced risk of certain cancers. The n-3 fatty acid DHA is an important structural lipid in the brain, central nervous system and retina.Diet rich in these have helped to avoid the same. It is vital that nutrition scientists communicate effectively for a proper diet in order to either reduce or correct such diseases. Replacing saturated fat with carbohydrates, has not been found to benefit for CVD risks. Supplementation of polyunsaturated fats in place of saturated fats or carbohydrates has been found to reduce the risk. Much research has been done to show the health benefits of dietary patterns (A healthy approach to dietary fats: understanding the science and taking action to reduce consumer confusion (Ann et al., 2017).

Requirement and metabolism of Carbohydrates: The primary functions of carbohydrates is to provide energy to the body. The body cells convert carbohydrates into ATP through cellular respiration. This provides energy, stores energy and builds up macromolecules and helps in lipid metabolism. The carbohydrates in the foods we eat are digested and broken down into glucose before entering the bloodstream.The energy is stored in the form of glycogen. Majority lies in the muscles and liver. Usually the liver contains approximately 100 grams of glycogen. These stored glucose molecules can be released into the blood to provide energy throughout the body and helps in maintaining the normal blood sugar levels between meals.Unlike liver glycogen, the glycogen in the muscles are used by the muscle cells. When glucose from carbohydrates gets exausted then muscles get broken down into amino acids and gets converted into glucose or other compounds to generate energy. However, loss of muscle mass has been associated with poor health and a higher risk of death. Therefore, it is important to consume some carbohydrates in the diet to prevent the starvation related loss of muscle mass. Fibres are also very important in diet which passes through the body undigested and are either soluble or insoluble.Soluble fibres can be obtained from oats, legumes and the inner part of fruits and some vegetables. They absorb water and form a gel-like substance. As soluble fibre passes through the small intestine, it binds to bile acids and prevents them from being reabsorbed. To make more bile acids, the liver uses cholesterol that otherwise stays in the blood.This increases the bulk of faeces and helps to make bowel movements easier as it softens. This reduces strain and pain and constipation is also reduced. The insoluble fiber helps in alleviating constipation by adding bulk to the stool and makes the movement faster through the digestive tract and protects against digestive tract diseases. This type of fiber is found in whole grains and the skins and seeds of fruits and vegetables. It has also shown significant reductions in fasting blood sugar. It showed positive results in people with type 2 diabetes. In the body the stored energy is mainly in the form of triglyceride molecules in fat tissue. In times of starvation or very low-carb diets, the main fuel source is the utilization of ketones that are formed from the breakdown of fatty acids. It is the process of Ketosis when the body produces large amounts of ketones to use for energy and reduces the amount of muscles that needs to be broken down to be converted to glucose for energy(Keith Pearson, 2017, UNICEF). As a person consumes carbohydrates, the digestive system breaks them into glucose which enters the blood and raises the levels of blood sugar or glucose. When blood glucose levels rise, beta cells in the pancreas release insulin.Insulin is a hormone that allows the cells to absorb blood sugar for energy or storage. As the cells absorb the blood sugar, blood sugar levels starts to drop.When blood sugar levels drop below a certain point, alpha cells in the pancreas release glucagon. Glucagon is a hormone that makes the liver release glycogen, a sugar stored in the liver. Both, insulin and glucagon help in maintaining levels of blood glucose in the cells. Insulin brings excess blood glucose level down, while glucagon brings levels back up if it is low. There can be insulin resistance when the blood glucose levels rise very fast. Then, the cells, do not respond to insulin. As the pancreas produce high levels of insulin continuously, the beta cells in the pancreas wear out which drops the insulin production. This can lead to many health problems, like high blood pressure, high triglycerides, low levels of good HDL cholesterol, weight gain and is linked to type 2 diabetes.Saccharide is another word for carbohydrate. Foods high in carbohydrates include bread, pasta, beans, potatoes, rice and cereals.One gram of carbohydrate contains approximately 4 kilocalories, High glycemic index (GI) carbohydrates quickly enter the bloodstream as glucose. Switching to a low-GI diet improves the chance of a healthy weight and lifestyle. Carbohydrate-rich foods are bread, pasta, beans, potatoes, bran, rice and cereals. They have a high starch content. Sae-Eaw et al., (2007) studied the acceptance and purchase intent of US consumers for non wheat rice butter cakes. The United States (U.S.) Dietary Guidelines (2015-2020) recommend obtaining 45 to 65 percent of energy needs from carbohydrates, and a maximum of 10 percent should come from simple carbohydrates. Simple carbohydrates as sugar, increase the risk of obesity.It usually happens due to the reduction in physical activity, a higher consumption of junk food, a higher consumption of food additives, such as coloring, taste enhancers and artificial emulsifiers. The developing countries, like China, India, Brazil and Mexico, have seen an upsurge in obesity with the change in living standards and dietary habits. It was found that earlier people used to eat more natural produce and less junk food, were more physically active and slept for atleast eight hours each night. The carbohydrates present in fruits and vegetables, legumes and whole grains enter the bloodstream slowly as compared to the carbohydrates in processed foods.The carbohydrates in

Table 2: Requirement of Energy, Fat and Carbohydrates based on age and Gender

Group	Category/Age	Body Weight (Kg)	Net Energy cal/d)	Protein (g/d)	Visible Fat /d)	Cal-cium (g/d)	Iron (mg/d)	Zinc (mg/d)	Mag-nesium (mg/d)
Men	Sedentary work	60	2320	60.0	25	600	17	12	340
	Moderate work		2730		30				
	Heavy wok		3490		40				
Women	Sedentary work	55	1900	55	20	600	21	10	310
	Moderate work		2230		25			12	
	Heavy work		2850	78	30	1200	35		
	Pregnant		+350		30		21		
	Lactating 0-6m		+600	74	30	1200			
	6-12		+520	68	30				
Infants	0-6 months	5.4	92 kcal/kg/d	1.69g/kg/d	-	500	46µg/kg/d	- -	30
	6-12 months	8.4	80 kcal/kg/d	1.16g/kg/d	19		05	-	45
Children	1-3 years	12.9	1060	16.7	17	600	09	5	50
	4-6 years	18.0	1350	20.1	25		13	7	70
	7-9 Years	25.1	1690	29.5	30		16	8	100
Boys	10-12 years	34.3	2190	39.9	35	800	21	9	120
Girls	10-12 years	35.0	2010	40.4	35	800	27	9	160
Boys	13-15 years	47.6	2750	54.3	45	800	32	11	165
Girls	13-15 years	46.6	2330	51.9	40	800	27	11	210
Boys	16-17 years	55.4	3020	61.5	50	800	28	12	195
Girls	16-17 years	52.1	2440	55.5	35	800	26	12	235

Source: Nutrient Requirements and Recommended Dietary Allowances for Indians (ICMR 2010)

junk and processed foods and drinks can cause a person to feel hungry soon as they cause the glucose and insulin production levels to rise quickly.

Requirement of Macronutrients: In Table (2) it can be observed the requirement of Energy, Fat and Carbohydrates based on age and gender.

Carbohydrates that raise the blood sugar quickly are high on the glycemic index (GI), while those with a gentler effect on blood sugar levels have a lower GI score.It is the milling and grinding process that increases the GI of a food, eliminates other nutrients as minerals, vitamins and dietary fibers and starch is mainly left.For a low-GI diet, eating of unrefined foods like oats, barley or bran for breakfast is beneficial. Under the less refined products commonly used are wholegrain bread, brown rice, fresh fruit and vegetables. Consumption of fresh whole fruit instead of juice, whole grain pasta, salads and raw vegetables are found to be good. The recommendations for quantities consumed of different food groups in volumes can be summarized for various population groups as given in Table (3).

Table 3

a) Children and adolescents

Group	1-5 years	6-14 years	14-18 years	14-18 years
M/F	M/F		Males	Females
Whole grains	0.5 c*	1.0 c	1.5 c	1.0 c
Refined grains	2.0 c	3.0 c	4.5 c	4.0 c
Fruits	2.0 c	3.0 c	5.0 c	4.0 c
Vegetables	1.5 c	2.5 c	4.5 c	4.0 c
Meats	60 g	100 g	130 g	100 g
Legumes	0.5 c	1.0 c	1.0 c	1.0 c
Milk; Dairy	0.3 c	0.5 c	1.0 c	0.5 c

c = cups. 1 cup grains = 1 slice of bread; ½ cup cooked cereal rice or pasta or 1 cup ready to eat cereal flakes.

Fruits and vegetables: 1 cup = 1 cup raw fruits or vegetables or ½ cup chopped, cooked or canned fruit or vegetable, ¾ cup of fruit or vegetables juice.

Milk and dairy: 1 cup equivalent=1.5 oz (42.5g) natural cheese (e.g cheddar); or 2 oz (56.7g) processed cheese (e.g. American).

b) Adults and Elderly:

Group	19-70	19-70	>70
Males	Females		M/F
Grains			
Whole	1.0 c	1.0 c	1.0 c
Refined	3.5 c	2.5 c	2.5 c
Fruits	4.0 c	3.5 c	3.0 c
Vegetables	3.0 c	3.0 c	2.5 c
Meats	100 g	80 g	75 g
Legumes	1.0 g	0.5 c	1.0 c
Milk; Dairy	0.5 c	0.5 c	0.5 c

c= cups. 1 cup grains = 1 slice of bread; ½ cup cooked cereal rice or pasta or 1 cup ready to eat cereal flakes.

Fruits and vegetables: 1 cup = 1 cup raw fruits or vegetables or ½ cup chopped, cooked or canned fruit or vegetable, ¾ cup of fruit or vegetables juice.

Milk and dairy: 1 cup equivalent=1.5 oz (42.5g) natural cheese (e.g cheddar); or 2 oz (56.7g) processed cheese (e.g. American).

Table 4: Gives the Amount of different Fatty acids required by the Body

Dietary Factor	
Total fat	15-30%
Saturated fatty acids	<10%
Polyunsaturated fatty acids	6-10%
n-6 Polyunsaturated fatty acids (PUFAs)	5-8%
n-3 Polyunsaturated fatty acids (PUFAs)	1-2%
Trans fatty acids	<1%
Monounsaturated fatty acids (MUFAs)	By difference
Total carbohydrates	55-75%
Free Sugars	<10%
Protein	10-15%
Cholesterol	<300mg/per day
Sodium Chloride (Sodium)	<5 g per day (<2 g per day)
Fruits and vegetables	>= 400 g per day

Source: Ranges of population nutrient intake goals *Source*: Diet, Nutrition and Prevention of Chronic Diseases; WHO report series 916; Geneva 2003

Requirement and working of Minerals: These are inorganic nutrients which are required in small amounts from less than 1 to 2500 mg per day. These are used in the preparation of hormones and to maintain a normal heartbeat. Large amounts of calcium required for the construction and maintenance of bone and normal function of nerves and muscles are met by them. Phosphorus is an important constituent of adenosine triphosphate (ATP) and nucleic acid. It is essential for bones, tooth formation and for RBC function. Iron is also an important component of the cytochromes that helps in cellular respiration. Magnesium, copper, selenium, zinc, iron, manganese and molybdenum are important co-factors used for enzyme reaction and many biochemical reactions. For thyroid hormones, iodine is required. For the maintenance of osmotic balance between cells and the interstitial fluid many minerals like sodium, potassium and chlorine are needed. Magnesium is an important component of chlorophyll in plants. There is a direct link between nutrition and diseases and also nutrition and drug metabolism. Each body has a different requirement of minerals and therefore, intake of minerals should be as per the requirement by the body. If taken in excessive quantity it can upset the homeostatic balance and could cause toxic side effects. Excess intake of sodium is associated with high blood pressure and excess of iron could cause liver damage. Change in mineral content can alter the delicate balance in body functions. It is therefore very important to consume a balanced diet.The tables below (Table 5, 6) list minerals, what they do in the body (their functions), and their sources in food. "A mineral is an element or chemical compound that is normally crystalline and that has been formed as a result of geological processes" (Nickel, 1995).

Table 5

Macrominerals:Major minerals

Mineral	Function	Sources
Sodium	Needed for proper fluid balance, nerve transmission, and muscle contraction	Table salt, soy sauce; large amounts in processed foods; small amounts in milk, breads, vegetables, and unprocessed meats
Chloride	Needed for proper fluid balance, stomach acid	Table salt, soy sauce; large amounts in processed foods; small amounts in milk, meats, breads, and vegetables
Potassium	Needed for proper fluid balance, nerve transmission, and muscle contraction	Meats, milk, fresh fruits and vegetables, whole grains, legumes
Calcium	Important for healthy bones and teeth; helps muscles relax and contract; important in nerve functioning, blood clotting, blood pressure regulation, immune system health	Milk and milk products; canned fish with bones (salmon, sardines); fortified tofu and fortified soy milk; greens (broccoli, mustard greens); legumes
Phosphorus	Important for healthy bones and teeth; found in every cell; part of the system that maintains acid-base balance	Meat, fish, poultry, eggs, milk, processed foods (including soda pop)
Magnesium	Found in bones; needed for making protein, muscle contraction, nerve transmission, immune system health	Nuts and seeds; legumes; leafy, green vegetables; seafood; chocolate; artichokes; "hard" drinking water
Sulfur	Found in protein molecules	Occurs in foods as part of protein: meats, poultry, fish, eggs, milk, legumes, nuts

Source: https://myhealth.alberta.ca

Trace minerals (microminerals)

The body needs trace minerals in very small amounts. Note that iron is considered to be a trace mineral, although the amount needed is somewhat more than for other microminerals. Natalie (2018) has discussed on the health benefits and nutritional value of spinach.

Table 6

Trace minerals

Mineral	Function	Sources
Iron	Part of a molecule (hemoglobin) found in red blood cells that carries oxygen in the body; needed for energy metabolism	Organ meats; red meats; fish; poultry; shellfish (especially clams); egg yolks; legumes; dried fruits; dark, leafy greens; iron-enriched breads and cereals; and fortified cereals
Zinc	Part of many enzymes; needed for making protein and genetic material; has a function in taste perception, wound healing, normal fetal development, production of sperm, normal growth and sexual maturation, immune system health	Meats, fish, poultry, leavened whole grains, vegetables
Iodine	Found in thyroid hormone, which helps regulate growth, development, and metabolism	Seafood, foods grown in iodine-rich soil, iodized salt, bread, dairy products
Selenium	Antioxidant	Meats, seafood, grains
Copper	Part of many enzymes; needed for iron metabolism	Legumes, nuts and seeds, whole grains, organ meats, drinking water
Manganese	Part of many enzymes	Widespread in foods, especially plant foods
Fluoride	Involved in formation of bones and teeth; helps prevent tooth decay	Drinking water (either fluoridated or naturally containing fluoride), fish, and most teas
Chromium	Works closely with insulin to regulate blood sugar (glucose) levels	Unrefined foods, especially liver, brewer's yeast, whole grains, nuts, cheeses
Molybdenum	Part of some enzymes	Legumes; breads and grains; leafy greens; leafy, green vegetables; milk; liver

Source: Kathleen Romito, Rhonda O'Brien, 2018. Healthwise Staff.Medical Review.

Other trace nutrients known to be essential in tiny amounts include nickel, silicon, vanadium and cobalt.The essential minerals helps the body to stay healthy. However, the body cannot make them. The 16 essential minerals are calcium, phosphorus, potassium, sulfur, sodium, chloride, magnesium, iron, zinc, copper, manganese, iodine, selenium, molybdenum, chromium and fluoride. Most of them are present in Brazil nuts, cashew nuts, cheese, eggs, milk, chicken, lean meat, liver, garlic, onion, green vegetables.mackerel, salmon, tuna, sunflower seeds and whole wheat bread.

Calcium is the most abundant mineral in the body. Around 99 percent of calcium is stored in the bones and teeth and supports their structure and function. The remaining one percent is used to support metabolic functions. It can be obtained from milk, yogurt, cheese, paneer, broccoli, chickpeas and fish. It is the most important mineral in the human body for optimal bone health throughout the life. Adults require a minimum of 1000 mg/day of Ca. Calcium Source is present in many forms. There are many who take ground oyster shell or Calcium Carbonate, where the bioavailability is very poor. The complex organic acid salts like citrate, lactate, gluconate, fumarate and malate have more bioavailability. Experiments have shown that calcium, along with vitamin D, may have many more benefits beyond bone health. These are protecting against cancer, diabetes and high blood pressure. Dietary reference intakes for calcium and vitamin D. Washington, DC: National Academy Press, 2010). Adequate Intake (AI) Source: National Institutes of Health; Office of Dietary Supplements ("New PubMed® subset - "Dietary Supplements" (U.S.Government content and in the Public Domain).

There is also data that indicates that Ca plays a role in preventing colorectal cancer suggesting that there may be a protective effect (Committee to Review Dietary Reference Intakes for Vitamin D and Calcium, Food and Nutrition Board, Institute of Medicine (National Academy Press, 2010).Several studies have found that higher intake of calcium from foods (low-fat dairy Sources) and/ or supplements are associated with a decreased risk of colon cancer (Slattery et al., 1999). In a follow-up study, supplementation with calcium carbonate led to reductions in the risk of adenoma (a nonmalignant tumor) in the colon, a precursor to cancer even as long as 5 years after the subjects stopped taking the supplement (Baron et al., 1999). In two large prospective epidemiological trials, men and women who consumed 700–800 mg per day of calcium had a 40%–50% lower risk of developing left-side colon cancer (Wu et al., 2002). But other observational studies have found it inconclusive. In the Women's Health Initiative, a clinical trial which involved 36, 282 postmenopausal women, it was found that daily supplementation with 1, 000 mg of calcium and 400 International Units (IU) of vitamin D3 for 7 years produced no significant differences in the risk of invasive colorectal cancer compared to placebo (Wactawski-Wende et al., 2006). The authors of a Cochrane systematic review concluded that calcium supplementation might moderately help in the prevention of colorectal adenomas. However, there is not enough evidence to recommend routine use of calcium supplements to prevent colorectal cancer. Given the long latency period for colon cancer development, long-term studies are needed to fully understand whether calcium intakes affect colorectal cancer risk. Not all calcium consumed is actually absorbed in the gut. Only 30% of the calcium in foods is absorbed by the human body, however, it depends on the type of food one consumes (Dietary Reference Intakes for Calcium and Vitamin D. Washington, DC: National Academy Press, 2010). As the calcium intake increases the efficiency of absorption decreases (Intakes for Calcium and Vitamin D. Washington, DC: National Academy Press, 2010). It has been found out that the absorption is age and life stage dependent. Total calcium absorption is high (60%) in infants and young children, where it is needed in large amounts for the building of bones (National Institutes of Health. According to NIH Consensus Statement (1994) absorption decreases to 15%–20% in adulthood (though it increases during pregnancy) and continues to decrease with age as compared to younger age. Therefore, the recommended calcium intake is higher for females older than 50 years and for both males and females older than 70 years. It has been observed that the presence of vitamin D improves absorption of calcium. Vitamin D is obtained from food and produced by skin when exposed to sunlight of sufficient intensity and time. There are other components in food like phytic acid and oxalic acid, which binds to calcium and can inhibit its absorption. Foods with high levels of oxalic acid include spinach, collard greens, sweet potatoes, rhubarb and beans. Among the foods high in phytic acid are fiber-containing whole-grain products like wheat bran, beans, seeds, nuts and soy isolates. The extent to which these compounds affect the absorption of calcium is different. Recently, it has been found out that eating spinach and milk at the same time reduces absorption of the calcium in milk. In contrast, wheat products (with the exception of wheat bran) do not appear to lower calcium absorption. For people who eat a variety of foods, these interactions do not have much nutritional consequence. High levels of calcium in the blood (hypercalcemia) can cause renal insufficiency, vascular and soft tissue calcification, hypercalciuria (high levels of calcium in the urine) and kidney stones. High calcium intake can cause constipation. It might also interfere in the absorption of iron and zinc, though this effect is not well established (Dietary Reference Intakes for Calcium and Vitamin D. Washington, DC: National Academy Press, 2010). In adults, too much calcium (from dietary supplements but not food) might increase the risk of kidney stones. Some studies have shown that people who consume high amounts of calcium can have higher risks of prostate cancer and heart disease, but, more research is needed to clarify these possible links (National Institute of Health, Office of Dietary Supplements). Herrera et al., (2007) have studied the school based mentoring to children on health and food.

Surveys show that some older women in the US probably get amounts above the upper limit since the use of supplements is common among these women. Life Stage Upper Limit Birth to 6 months 1, 000 mg, infants 7–12 months 1, 500 mg, children 1–8 years 2, 500 mg, children 9–18 years 3, 000 mg, adults 19–50 years 2, 500 mg, adults 51 years and older 2, 000 mg, pregnant and breastfeeding teens 3, 000 mg, pregnant and breastfeeding adults 2, 500 mg. Getting too much calcium from foods is very rare, excess intakes are more likely to be caused by the use of calcium supplements. In older women, over consumption of calcium is also linked to CVD (cardiovascular disease) (Diane et al. , 2013). Calcium helps to reduce cardiovascular disease (CVD) risk by decreasing intestinal absorption of lipids, increasing lipid excretion, lowering cholesterol levels in the blood and promoting a calcium influx into the cells. The effect of dietary calcium has different effects on the cardiovascular system than supplemental calcium is not very clear. In the Iowa Women's Health Study, higher calcium intake from the diet and/or from the supplements has been associated with reduced ischemic heart disease mortality in postmenopausal women (Bostick et al., 1999). However, it is still not clear that Ca supplementation can lower cardiovascular disease occurrence or not. Jost Chemical Co. in USA has come out with a variety of organic calcium salts that have the bioavailability and these have been successfully used as Ca supplementation. These products are produced under Q7 GMP regulations and are fit for food, dietary supplement and pharmaceutical uses. Another Ca salt produced is the citrate malate which has shown high bioavailability. The most water soluble Ca salts are lactate (3.1 gm in 100 ml at RT), fumarate (2 gm in 100ml at RT) and malate (0.8gm in 100ml at RT). Other salts are almost insoluble with much less than 1gm solubility. The insoluble salts can be used in liquids where a suspension is possible. There is a need to have products that contain the correct ratio of Ca:Mg of 2:1 as the daily requirement is in this ratio. There is a strong tendency for mixtures of Ca- and Mg- citrates to set up hard in solution. They become cement-like. This is a natural reaction that occurs often in water suspensions. Chloride helps the body salt (sodium chloride) which maintains the proper balance of water. Good sources are tomatoes, lettuce and pickles. Magnesium helps in muscle and nerve function, controls blood glucose levels, regulates blood pressure and makes DNA, proteins and bones. It can be obtained from legumes, seeds, nuts, whole grains, fortified cereal, milk, yogurt and green leafy vegetables like spinach. Phosphorus works with calcium to help build strong bones and teeth. It also plays an important role in energy metabolism, helps to filter out the waste from the kidneys. Also helps in the growth, maintenance and repair of the tissues and cells. It is best obtained from milk, dried fruit, whole grains and protein-rich foods that include meat poultry, eggs, fish, nuts and legumes. Potassium connects between nerves and muscles and helps the movement of nutrients into the cells.It is found in blackberries, grapes, oranges, grapefruit, bananas, leafy greens like spinach, carrots and potatoes. Sodium can be obtained from cheese, salted nuts and many processed foods like bread, cereal, pickles and sauces. Sulfur is required for body's connective tissues, including tendons, ligaments and cartilage. It can be obtained from meat, fish, poultry, eggs, legumes, garlic, onions and asparagus.

Trace mineral chromium helps in converting the fat, protein and carbohydrates into energy and plays a role in breaking down insulin. Whole grains, oatmeal, mushrooms, broccoli, potatoes, garlic, basil and grape juice. The role of copper is in forming red blood cells, keeping the immunity, blood vessels nerves and bones healthy. It can be obtained from shellfish, liver meat, whole grains, nuts, beans, potatoes, dried fruit and green leafy vegetables.Fluoride helps to reduce the tooth decay and maintain the bone structure. It is present in fluoridated water, tea and most of the seafoods. Iodine helps in thyroid hormones, development of the brain and bone development during pregnancy and infancy can be obtained from iodized salt, seafood, milk, yogurt, cheese eggs, bread, cereals and some fruits and vegetables. The amount depends on the quality of the soil and fertilizer used for growing. Iron helps in making different proteins, including hemoglobin, a protein that carries oxygen to various tissues.It can be obtained from lean meat, seafood and poultry, iron fortified bread and cereals, legumes like lentils, white and myoglobin.Iron is used by the body for growth and development, and to make some connective tissue and hormones. It can be obtained from beans, kidney beans, peas, nuts and some dried fruits like raisins. Manganese helps in connective tissue, bone formation, calcium absorption, digestion of fat and carbohydrate, control of blood sugar, brain and nerve function. It can be obtained from whole grains nuts, seeds, legumes and pineapple. Manganese is a vital mineral as it helps in mental function. A deficiency can adversely affect many bio-chemical processes in the body but may be difficult to determine. The human body contains about 15 to 20 milligrams of it. Adults require only about 2 mg per day and is present in the bones and in tissues like the pancreas, kidneys, liver adrenal glands and pituitary glands. Due to the highly processed modern diet 1/3 of the North American population do not get enough of this mineral. It activates the enzymes required to digest and synthesize fatty acids and cholesterol, metabolizes carbohydrates and proteins, helps to produce energy activates enzymes that enables the utilization of vitamin B_1, biotin, choline ascorbic acid and vitamin E. It helps in the normal growth of bone and cartilage and keeps them strong and healthy. It helps in the formation of collagen in skin cells, which is required for healing of wounds. It has an important role in glucose metabolism which helps in regulating a proper balance between blood and sugar. It is required in the preparation of manganese superoxide dismutase (MnSOD), which is an antioxidant that protects the cells from free radical damage and helps to maintain a healthy immune system. It also helps in the production of sex hormones, fertility, reproductive health, lactation and in the maintainance of healthy nerves. It supports an optimal function of the thyroid gland and thyroxine production. It is essential for proper iron metabolism and helps in the prevention of anemia. Mitochondria consumes more than 90% of the oxygen used by cells and are subjected to oxidative stress. The superoxide radical is one of the reactive oxygen species produced in mitochondria during ATP synthesis. MnSOD catalyzes the conversion of superoxide radicals to hydrogen peroxide, which can be reduced to water by other antioxidant enzymes (Leach and Harris; 1997).

For magnesium the recommended daily intake (RDI) is 400 mg/day for adult males with slightly lower amounts for women and children. It is required in almost 300 enzyme reactions in the human body. It helps the body to produce energy, protein synthesis, muscle and nerve function, blood glucose control, heart function and in maintaining the blood pressure by controlling the level of glucose in the blood (Djoussé & Gaziano (2009), Aranceta & Pérez (2012). Therefore, it has been found that heart patients and diabetic's require supplementation of Mg. It regulates the secretion and function of insulin and is necessary for insulin to open the cell membranes for glucose helping the body to digest, absorb and utilize proteins, fats and carbohydrates. Researchers have shown that lower levels of magnesium in the blood leads to metabolic malfunctions such as diabetes and impaired kidney function that is associated with diabetes.It has also been found that the higher levels of magnesium in the urine and less of magnesium intake causes high blood sugar levels. It helps to convert CO_2 and H_2O into sugar in chlorophyll and therefore, plants are a good source of Mg. The green leafy vegetables are particularly good but it is lost during processing. When the soil is depleted of Mg due to constant farming it reflects in the produce. Molybdenum is the micronutrient which helps in the making of many enzymes that are required during the body cellular metabolism. It is present in nuts, legumes, grains and milk. Selenium plays a critical role in DNA production, thyroid hormone metabolism, reproduction and protecting the body from damage caused by infection or free radicals. It is present in many seafoods, meat, poultry, eggs, bread, cereals and dairy products. Zinc is a mineral element which serves as a cofactor in as many as 300 enzymes required for human metabolism. It is required for the regulation of synthesis and degradation of nucleic acids, proteins, lipids and carbohydrates. Zinc also plays a vital role in maintaining cellular integrity and immune system. The EAR of zinc for non-pregnant and non-lactating women is 2960 μg/g, while it is 1390 μg/g per day in case of children of 4-6 yr(Andersson et al, 2017). The clinical symptoms of severe zinc deficiency includes; retardation in growth associated with delayed sexual and bone maturation. Loss of appetite, impaired immune function, increased susceptibility to infections such as diarrhoea besides skin lesion also occurs as a result of zinc deficiency (Prasad, 1996). Zinc is needed by the body during pregnancy, infancy and childhood as it promotes growth and development. It also helps in making various proteins, DNA, healing of the wounds, boosting the immune system and supports for taste buds and smell. It is present in oyster mushrooms, crab, lobster, red meat, poultry, nuts, whole grains, beans and fortified cereals. In order to have normal functions in the body it is vital to eat a balanced diet. According to the statistics, billions of people around the world have vitamin and mineral deficiencies and therefore in such cases supplementation becomes necessary (Conseil Supérieur de la Santé (2009). In India, 74 percent of the population suffers from iron deficiency called as anemia (Nordic Council of Ministers, 2012). Indians are prone to iodine deficiency. This could be overcome if the soil is rich in the nutrient in which the crop grows. World Cancer Research Fund (WCRF) / American Institute for Cancer Research (AICR; 2007) Institute of Medicine (IOM) has extensive data that gives the general Daily Reference Intake (DRI) for every human age group and gender. Iron is a mineral element which humans require for the proper functioning of muscle and brain tissues. It serves as the transport medium for electrons within cells and also carries oxygen from the lungs to various tissues by red blood cell haemoglobin. It is an integral part of important enzyme systems in various tissues. Estimated average requirement (EAR) of iron for non-pregnant and non-lactating women is 1460 μg/g, while it is 500 μg/g per day for children of 4-6 yr (Andersson et al, 2017). The occurrence of anaemia is the most common characteristics of iron deficiency in human. It affects overall growth and development, besides causing mild mental retardation and goitre thereby severely affecting the work productivity (Scrimshaw, 1984).The upper limit is the amount over which the mineral could be toxic. Generally the large mineral complex are more bioavailable than smaller molecules.

Copper (Cu) Copper is a trace element and is essential for the human body for proper functions. Copper is present in all body tissues and plays a role in the formation of connective tissue. It is helpful for the normal functioning of muscles, the immune and nervous systems. The human body requires copper for normal growth and health. Copper, along with iron, is a critical component in the formation of red blood cells. Copper also influences the functioning of the heart and arteries, helps in the prevention of bone defects like osteoporosis and osteoarthritis. It promotes healthy connective tissues (hair, skin, nails, tendons, ligaments and blood vessels (Feleke et al., 2018). Recommended Dietary Allowances (RDAs) for Copper varies with age, gender, female pregnancy and lactation.The common requirements are from birth to 6 months, 200mcg; 7–12 months, 220 mcg; 1–3 years mcg, 340 340 mcg; 4–8 years mcg; 440 440 mcg; 9–13 years 700 mcg, 14–18 years, 890 mcg; 19–30 years, 900 mcg; 31–50 years, 900 mcg; 51+ years 900 mcg. Adequate Intake (AI): (Organic Facts on Internet): Arthritis: The use of copper bracelets is common as it is in touch with the body. Copper can also work as a home remedy for arthritis. Traditionally water is being stored in a copper container overnight to accumulate copper traces. This water has benefits in strengthening the muscular system. Many people make it a habit to drink the water in the morning from a copper vessel. It is very important to include this mineral in balanced levels in our regular diets. It is also helpful in the protection of the skeletal, nervous and cardiovascular systems. Copper deficiency, stops the normal and healthy growth of organs and tissues and the proper oxygenation from an ample red blood cell. Copper deficiencies are seen in many third world countries and are reflected in the number of birth and growth defects in children of those nations.

For the proper pigmentation of hair and good eye sight copper is vital. It is a component of the natural dark pigment, melanin, which imparts coloration to the skin, hair and eyes. Melanin pigment is produced by melanocytes only in the presence of the cuproenzyme called tyrosinase, which is derived from copper. Intake of copper supplements helps in protecting against graying of the hair and in maintaining the color of the eyes. The deficiency occurs in people who are severely undernourished or who have chronic diarrhea. Improper absorption of the nutrient may cause Crohn's disease, it can lead to copper deficiency. Deficiency causes bleeding under the skin, damage of the blood vessels, hair loss, pale skin and an enlarged heart. Symptoms include fatigue and more prone to infections. Good food sources include vegetables, legumes beans, nuts, seeds, mushrooms, shellfish (especially cooked oysters), avocado and whole grains. If consumed in large amounts it can be toxic. In case of acute poisoning it can lead to nausea, vomiting, diarrhea, kidney damage, anemia and death. The toxic amount for adults is 10, 000 micrograms. It is advisable not to take more than the prescribed amount everyday. For younger people smaller amounts become toxic. Wilson's disease is an inherited disorder in which excessive amounts of copper accumulate in the liver or brain. If a person has Wilson's disease or a family history they should not take supplements containing copper and get regular testing to monitor the copper levels. www. wilsonsdisease.org.

The levels of copper in the body gets affected with the levels of boron, vitamin C, selenium, manganese and molybdenum. The levels of zinc can lower the storage of copper and is therefore recommended to take supplemental copper (1 : 10). The health benefits of copper relate to its anti-inflammatory actions that assist in reducing the symptoms of arthritis (Sandra, Educational programs of the Kentucky Cooperative Extension Service. Issued 03-96; FN-SSB. 028LG2012. Organic facts on internet : Arthritis (2020) has discussed benefits of copper).

Iron (Fe) is naturally present in many foods and is available as a dietary supplement. Iron is an essential component of hemoglobin. It is an erythrocyte protein that carries the oxygen from the lungs to the various tissues of the body. It is a component of myoglobin, that provides oxygen to the muscles. It is also necessary for growth, development, normal cellular functions, synthesis of some hormones and connective tissue (National Institutes of Health; Office of Dietary Supplements, Jost chem.co.).The remaining iron is stored in the form of ferritin or hemosiderin (a degradation product of ferritin) in the liver, spleen, bone marrow and in myoglobin in muscle tissue. Humans typically lose only small amounts of iron in urine, feces, the gastrointestinal tract and skin. During menstruation, women lose a lot of blood. Iron-deficiency anemia (IDA), occurs at the time when the blood hemoglobin concentrations are low. The most commonly used measures to screen patients for iron deficiency are hemoglobin and hematocrit tests, even though they are neither sensitive nor specific. Hemoglobin concentrations lower than 13 g/dL in men and 12 g/dL in women are the indicators of IDA. The intake recommendations for iron and other nutrients are provided in the Dietary Reference Intakes (DRIs) which is prepared by the Food and Nutrition Board (FNB) at the Institute of Medicine (IOM) of the National Academy of Sciences. DRI is the general term for a set of reference values used for planning and assessing nutrient intakes of healthy people. These vary with age and gender.As per the Recommended Dietary Allowance (RDA): average daily level of intake sufficient to meet the nutrient requirements of nearly all (97%–98%) healthy individuals. The RDAs for vegetarians are 1.8 times higher than for people who eat meat. This is because heme iron from meat is more bioavailable than nonheme iron from plant-based foods. Meat, poultry and seafood increase the absorption of nonheme iron. For infants from birth to 6 months, the FNB established an Adequate intake AI for iron that is equivalent to the mean intake of iron in healthy, breastfed infants. Recommended Dietary Allowances (RDAs) for iron varies with age, gender, pregnancy and period of lactation. From birth to 6 months the requirement is 0.27 mg; 7–12 months, 11 mg; 1–3 years, 7 mg; 4–8 years, 10 mg; 9–13 years, 8 mg; 14–18 years, 11-27 mg; 19–50 years, 8–27 mg; 51+ years, 8 mg. As per National Institutes of Health; Office of Dietary Supplements Jost Chemical Co.; the adequate Intake (AI) have been prepared by the manufactures in several compendial grades of Fe which are suitable for foods or dietary supplements like ferric ammonium citrate, ferric ammonium sulfate, ferric phosphate, ferrous gluconate dihydrate, ferrous lactate dihydrate, ferrous fumarate anhydrous, ferrous citrate dibasic monohydrate, and the ferric citrate x-hydrate. They have also manufactured encapsulated products of ferrous sulfate, ferrous fumarate and reduced iron.

Zinc is another mineral that must be absorbed in small amounts to keep humans healthy and is necessary for body growth, maturation, development, tissue repair and resistance to disease. Zinc is an important mineral for children and the elderly. It can be obtained from meats, specifically organ meats, poultry and seafood. Although rare, trace mineral deficiencies related to zinc can result in reduced growth in children, reduced resistance to infection in adults and delayed wound healing in people of all ages. Per day the body needs 15.0 milligrams of zinc. Zinc is an essential mineral that is naturally present in some foods. It is needed for cellular metabolism. It is required for the catalytic activity of around 100 enzymes and plays a role in immune function, protein synthesis, wound healing, DNA synthesis and cell division. It helps during pregnancy, childhood and adolescence. It is required for sensing taste and smell. A daily intake of zinc is required to maintain a steady state because the body has no specialized zinc storage system. (National Institutes of Health; Office of Dietary Supplements and references there-in). Zinc intake and other nutrients are available in the Dietary Reference Intakes (DRIs) developed by the Food and Nutrition Board (FNB) at the Institute of Medicine of the National Academies (formerly National Academy of Sciences). The deficiency characteristics are growth retardation, loss of appetite and impaired immune function. In severe cases, zinc deficiency causes hair loss, diarrhea, delayed sexual maturation, impotency, hypogonadism in males, lesions in the eye and skin. Weight loss, delayed healing of wounds, taste abnormalities and mental lethargy can occur. Being non-specific it needs to be tested by plasma or serum zinc levels. Zinc toxicity can occur both in acute and chronic forms. Acute adverse effects of high zinc intake include nausea, vomiting, loss of appetite, abdominal cramps, diarrhea and headache. One case report cited severe nausea and vomiting within 30 minutes of ingesting 4 g of zinc gluconate (570 mg elemental zinc). Intakes of 150–450 mg of zinc per day have been associated with such chronic effects as low copper status, altered iron function, reduced immune function and reduced levels of high-density lipoproteins. Eating a balanced diet is an essential part of staying healthy (Jenna Fletcher, 2018, Rev. Natalie Olsen). Brown et al., (2004) have written a technical document on the risk of Zn deficiency.

Foods to relieve rheumatoid arthritis

In a work done by Sandoiu (2017) has given a list of foods that have been proven to relieve the symptoms of rheumatoid arthritis in the long-term. Prunes are a great source of polyphenols and can help to ease inflammatory symptoms.Rheumatoid arthritis is an autoimmune disease that affects approximately 1.3 million adults in the United States.The fact that it is an autoimmune disorder means that the body does not recognize its own healthy cells and attacks them as though they are foreign. This causes inflammation in the joints and gets translated into stiffness and swelling. So-called disease-modifying antirheumatic drugs are normally the first line of treatment for this condition, for which there is no known cure at the moment. If a person living with rheumatoid arthritis does not react well to these drugs, so-called biological response modifiers, or "biologicals, " are the second-line treatment option.Whole grains, legumes, fruits and spices come under this category. Fruits include prunes, grapefruits, grapes, blueberries, bananas, pomegranate, mango, peaches and apples. Cereals include whole oatmeal, whole-wheat bread and whole rice. Among the whole grains are corn, rye, barley, millets and sorghum.

Spices including turmeric and ginger, olive oil, fish oil, green tea and yogurt are also among those listed as beneficial. These can reduce the level of cytokines or substances secreted by the immune cells that can cause inflammation in people with rheumatoid arthritis and could reduce oxidative stress, thereby improving the body's ability to Fight off toxins.Regular consumption of specific dietary fibers, vegetables, fruits and spices, as well as the elimination of components that cause inflammation and damage can help patients to manage the effects of rheumatoid arthritis. Incorporation of probiotics into the diet can also reduce the progression and symptoms of the disease.Nutraceuticals have an advantage over allopathic medicines as they have no side effects, they originate from natural sources and are economical.However, the general public must consult doctors and dieticians before eating or going through any diet program or food compounds.

Requirement of Vitamins: Fat-soluble vitamins get stored in the fat tissue of the body. These are vitamins A, D, E and K. The absorption of these vitamins is very easy by the body in the presence of dietary fat.The water-soluble vitamins are not stored in the body and leave the body through the urine. These are Vitamin A, Vitamin B1 (thiamine), Vitamin B2 (riboflavin), Vitamin B3 (niacin), Pantothenic acid (B5), Biotin (B7), Vitamin B6, Vitamin B12 (cyanocobalamin), Vitamin B9, Vitamin C, Vitamin D, Vitamin E, Vitamin K, Folate (folic acid) (Brian Krans in Healthline 2016, rev. by Natalie Butler). For the body functions these need to be consumed on a regular basis. Vitamin B12 is the only water-soluble vitamin that can be stored in the liver for many years. Choline and carnitine are also needed by the body.Each vitamin has a particular function in the body.A vitamin deficiency can cause health problems.Not eating enough fruits, vegetables, beans, lentils, whole grains and fortified dairy foods may increase the risk for health problems, including heart disease, cancer and poor bone health (osteoporosis).Vitamin A helps in the maintainance of healthy teeth, bones, soft tissue, mucous membranes, eyes and skin. Vitamin C, also known as ascorbic acid, is an antioxidant. It promotes healthy teeth and gums. It also helps the body to absorb iron and maintains healthy tissue. It is also essential in wound healing.

Vitamin A: Also called as retinol, vitamin A is essentially required for the normal functioning of the visual system, growth and development, maintenance of epithelial cell integrity, immune system and reproduction. The requirement of vitamin A in non-pregnant and non-lactating women is 500 μg/g, while it is 275 μg/g per day in case of children of 4-6 yr. Night blindness is the hallmark of an early symptom of vitamin A deficiency. Xerophthalmia and keratomalacia caused due to structural alterations of the conjunctiva and cornea may also follow. In severe deficiency, it may lead to irreversible blindness. Further, chances of anaemia, diarrhoea, measles, malaria and respiratory infections are also enhanced (West, 2000).

Vitamin B: Niacin is a B vitamin which helps in maintaining the skin and nerves healthy. It also has cholesterol-lowering effects at higher doses. Folate works with vitamin B_{12} to form red blood cells, for the production of DNA and controls tissue growth and cell function. If the folate levels are low they are linked to birth defects such as spina bifida. Many foods are now fortified with folic acid. Pantothenic acid is essential for the metabolism of food and helps in the production of hormones and cholesterol.Riboflavin (vitamin B_2) works with the other B vitamins for body growth and the production of red blood cells.Thiamine (vitamin B_1) helps the body cells change carbohydrates into energy. Getting enough carbohydrates is very important during pregnancy and breastfeeding. It is also essential for heart function and healthy nerve cells.Choline helps in normal functioning of the brain and nervous system. Lack of choline can cause swelling in liver.Carnitine helps the body to change fatty acids into energy. A balanced diet usually provides enough of these vitamins. Vitamin B_6 is also known as pyridoxine. It helps in the formation of red blood cells and maintains the function of the brain. This vitamin also plays an important role in the proteins that are part of many chemical reactions in the body. With more protein intake higher quantity of pyridoxine is required by the body.Vitamin B_{12}, is important for metabolism. It helps in the formation of red blood cells and also maintains the central nervous system. People older than 50 and some vegetarians usually require supplements to get enough B_{12}.

Vitamin C: It is also called ascorbic acid and required for metabolism and repair of various tissues such as skin, bone, teeth and cartilage. Vitamin C is also used to increase iron absorption from the gastrointestinal tract. The recommended dietary allowance for vitamin C in adult women is 60 mg per day, while it is 90 mg/day in adult men(FAO/WHO and World Health Organization; 2000). Deficiency leads to scurvy which is characterized by bleeding gums, bruising and poor wound healing in the teeth and is also associated with joint and muscle pains.

Vitamin D: is obtained from sunlight and is therefore also known as the "sunshine vitamin". 10 to 15 minutes of sunshine 3 times a week is enough to produce the body's requirement of vitamin D for most people in tropical areas or otherwise it is difficult to get vitamin D. It is not produced by food sources. Vitamin D helps the body to absorb calcium which is required for the normal development and maintenance of healthy teeth and bones. It also helps to maintain proper blood levels of calcium and phosphorus.

Vitamin E: is an antioxidant and is also known as tocopherol. It helps the body to form red blood cells and use vitamin K without which the blood does not coagulate. It is also important for bone health.Biotin is essential for the metabolism of proteins and carbohydrates and in the production of hormones and cholesterol.

Balancing carbohydrates, proteins and fats are important for better nutrition and health

The body uses calories from food for walking, thinking, breathing and other important functions.The average person needs to eat about 2, 000 calories per day to maintain their weight. According to Barlow (2007) an expert committee's recommendations regarding the prevention, assessment, and treatment of child and adolescent overweight and obesity has been given in a summary report. A person's specific daily calorie intake can vary depending on their age, gender and physical activity level. Men generally need more calories than women and people who exercise need more calories.The following examples of daily calorie intake are based on United States Department of Agriculture (USDA) guidelines: children age 2 to 8 years: 1, 000 to 1, 400 calories, girls age 9 to 13 years: 1, 400 to 1, 600 calories, boys age 9 to 13 years: 1, 600 to 2, 000 calories, active women age 14 to 30 years: 2, 400 calories, sedentary women age 14 to 30 years: 1, 800 to 2, 000 calories, active men age 14 to 30 years: 2, 800 to 3, 200 calories, sedentary men age 14 to 30 years: 2,000 to 2, 600 calories, active men and women over 30 years: 2, 000 to 3, 000 calories, sedentary men and women over 30 years: 1, 600 to 2, 400 calories. The source of the daily calories is very important as the number of calories consumed. The USDA defines empty calories as calories that come from sugars and solid fats like butter. According to the USDA, the Americans consume empty calories from bacon, sausages, cakes, cheese, cookies, doughnuts, energy drinks, fruit drinks, ice cream, pizza, sports drinks and sodas.A balanced diet is important as the organs and tissues need proper nutrition for working effectively and to combat diseases, infection, fatigue and poor performance. Children with a poor diet or malnutrition do not grow and develop as per their age and have low academic performance. Rising levels of obesity and diabetes in America are due to the effects of a poor diet and a lack of exercise. The Center for Science in the Public Interest has reported that 4 out of the top 10, leading causes of death in the United States are directly influenced by the diet. Among these the major ones are heart disease, cancer, stroke and diabetes.In a balanced diet the foods should be low in unnecessary fats and sugars and should be high in vitamins, minerals and other nutrients.The food groups which come under the essential parts of a balanced diet are as given under.

Fruits are a great source of nutrition and make tasty snacks. Seasonal fruits are fresh and provide a lot of nutrients. The sugar content is high in fruits. This sugar is natural and a better choice than other foods with added sugar. In a condition like diabetes, one must opt for low-sugar fruits such as melons, citrus and peaches. Vegetables are primary sources of essential vitamins and minerals. Dark, leafy greens generally contain the most nutrition and can be eaten at every meal. These include spinach, kale, green beans, broccoli and collard greens.According to the USDA, Americans consume refined white flour more than any other grain. Refined white flour has poor nutritional value because the hull of the grain or outer shell, is removed during the refining process. The hull contains the maximum nutrition. Whole grains are prepared using the entire grain, includes the hull and therefore, it provides much more nutrition. whole-grain products are good than white bread and pastas.

Meats and beans are the primary sources of protein, a nutrient which is very essential for the proper development of muscle and brain. For better proteins lean, low-fat meats like chicken, fish, pork and beef are good to eat. Removing the skin and trimming off any visible fat are easy ways to reduce the amount of fat and cholesterol in meats. The health and diet of the animal are important which influences the fatty acid profile of the meat and therefore, the animals fed on grass are good.

Dairy products provide calcium, vitamin D and other essential nutrients. However, they are also a major source of fat. Therefore, it is good to choose small portions of full-fat cheese and reduced-fat or fat-free milk and yogurt. Plant-based milks, made from flaxseeds, almonds or soy are typically fortified with calcium and other nutrients, making them excellent alternatives to dairy from cows.Oils should be used sparingly. Usage of low-fat and low-sugar products with almost no oil like salad dressing and mayonnaise are good. Use of olive oil in comparison to other fattier vegetable oils is better. Deep-fried foods contain many empty calories and should be avoided for better health. Table (7) gives the range of nutrient intake goals. Barrena and Sanchez (2009) studied the differences in consumer abstraction levels as a function of risk perception.

Table 7: Ranges of Nutrient intake goals

Dietary Factor	Goal (% of total energy unless otherwise stated)
Total fat	15-30%
Saturated fatty acids	<10%
Polyunsaturated fatty acids	6-10%
n-6 Polyunsaturated fatty acids (PUFAs)	5-8%
n-3 Polyunsaturated fatty acids (PUFAs)	1-2%
Trans fatty acids	<1%
Monounsaturated fatty acids (MUFAs)	By difference
Total carbohydrates	55-75%
Free Sugars	<10%
Protein	10-15%
Cholesterol	<300mg/per day
Sodium Chloride (Sodium)	<5 g per day (<2 g per day)
Fruits and vegetables	>= 400 g per day

(*Source*: Diet, Nutrition and Prevention of Chronic Diseases; WHO report series 916; Geneva 2003

Table 8: Water-soluble vitamins

Nutrient	Function	Sources
Thiamine (vitamin B1)	Part of an enzyme needed for energy metabolism; important to nerve function	Found in all nutritious foods in moderate amounts: pork, whole-grain or enriched breads and cereals, legumes, nuts and seeds
Riboflavin (vitamin B2)	Part of an enzyme needed for energy metabolism; important for normal vision and skin health	Milk and milk products; leafy green vegetables; whole-grain, enriched breads and cereals
Niacin (vitamin B3)	Part of an enzyme needed for energy metabolism; important for nervous system, digestive system, and skin health	Meat, poultry, fish, whole-grain or enriched breads and cereals, vegetables (especially mushrooms, asparagus, and leafy green vegetables), peanut butter
Pantothenic acid	Part of an enzyme needed for energy metabolism	Widespread in foods
Biotin	Part of an enzyme needed for energy metabolism	Widespread in foods; also produced in intestinal tract by bacteria
Pyridoxine (vitamin B6)	Part of an enzyme needed for protein metabolism; helps make red blood cells	Meat, fish, poultry, vegetables, fruits
Folic acid	Part of an enzyme needed for making DNA and new cells, especially red blood cells	Leafy green vegetables and legumes, seeds, orange juice, and liver; now added to most refined grains

Cobalamin (vitamin B_{12})	Part of an enzyme needed for making new cells; important to nerve function	Meat, poultry, fish, seafood, eggs, milk and milk products; not found in plant foods
Ascorbic acid (vitamin C)	Antioxidant; part of an enzyme needed for protein metabolism; important for immune system health; aids in iron absorption	Found only in fruits and vegetables, especially citrus fruits, vegetables in the cabbage family, cantaloupe, strawberries, peppers, tomatoes, potatoes, lettuce, papayas, mangoes, kiwifruit

Source: Nutritive Value of Foods. United States Department of Agriculture

Fat-soluble vitamins are stored in the cells of the body and are not easily excreted. They do not need to be consumed as often as water-soluble vitamins, although adequate amounts are needed. If you take too much of a fat-soluble vitamin, it could become toxic.A balanced diet usually provides enough fat-soluble vitamins.

Table 9: Fat-soluble vitamins

Fat-soluble vitamins

Nutrient	Function	Sources
Vitamin A (and its precursor*, beta-carotene) *A precursor is converted by the body to the vitamin.	Needed for vision, healthy skin and mucous membranes, bone and tooth growth, immune system health	Vitamin A from animal Sources (retinol): fortified milk, cheese, cream, butter, fortified margarine, eggs, liver Beta-carotene (from plant Sources): Leafy, dark green vegetables; dark orange fruits (apricots, cantaloupe) and vegetables (carrots, winter squash, sweet potatoes, pumpkin)
Vitamin D	Needed for proper absorption of calcium; stored in bones	Egg yolks, liver, fatty fish, fortified milk, fortified margarine. When exposed to sunlight, the skin can make vitamin D.
Vitamin E	Antioxidant; protects cell walls	Polyunsaturated plant oils (soybean, corn, cottonseed, safflower); leafy green vegetables; wheat germ; whole-grain products; liver; egg yolks; nuts and seeds
Vitamin K	Needed for proper blood clotting	Leafy green vegetables such as kale, collard greens, and spinach; green vegetables such as broccoli, Brussels sprouts, and asparagus; also produced in intestinal tract by bacteria

Source: All you need to know about fat-soluble vitamins (Fletcher Jenna , 2017, Rev. Debra Rose Wilson).

RDA for Fat soluble Vitamins

Vitamin A plays an important role in maintaining healthy vision. Without vitamin A, a person would suffer from severe vision issues.Vitamin A does not refer to one single vitamin but is a collection of compounds known as retinoids. Retinoids can be found both in the human body and in some food sources.The most important functions include vision and the immune system that are supported by it. It can be obtained through natural sources like fish liver oil, liver of animals and butter. There are animal sources which provide the active components that form retinols within the human body.Some plants also provide pro-vitamin A compounds known as carotenoid antioxidants. The most common is known as beta carotene, which is present in foods like kale, carrots and spinach.The RDA of vitamin A gets varied with age and gender. For infants (0–12 months): 400–500 micrograms (mcg), children aged 1–3: 300 mcg, children aged 4–8: 400 mcg, children aged 9–13: 600 mcg, adult women: 700 mcg and adult men: 900 mcg. Vitamin A deficiency is common in developing countries. Usually it is the vegetarians that have a higher risk of a deficiency as they do not get vitamin A through their normal diet. Symptoms of deficiency include hair loss, dry eyes, blindness, reduced immune function and skin issues. High levels may be toxic and is called as hypervitaminosis. People who take vitamin A supplements or eat copious amounts of fish liver oils have more risk. High levels of vitamin A are harmful to a growing fetus and could result in a baby with birth defects.An overdose, may create headaches and fatigue. The effectiveness of vitamin supplements can depend on the time they are taken.

Vitamin D is a collective term used to describe a collection of compounds. Collectively, these are often referred to as calciferol.There are two types found naturally which include vitamin D-3, present in animal fats and vitamin D-2, found in plants, like mushrooms.Once vitamin D is absorbed into the bloodstream, the liver and kidneys change calciferol into calcitriol, the biologically active form of vitamin D. It does bone maintenance and supports the immune system. When the body is exposed regularly to sun rays, then the body can absorb enough rays to produce vitamin D to function properly and then there is no need for supplements. Covering the body with sunscreen and clothing does not allow to absorb much vitamin D through sunlight. Good Sources of food for vitamin D are fish oil, fatty fish, mushrooms exposed to ultraviolet light and fortified dairy products.Recommended daily values of vitamin D varies with age. For infants (0–12 months): 10 mcg;1–70 years of age: 15 mcg;above age 70: 20 mcg. Deficiency of vitamin D is found very commonly. People who are at a higher risk of developing a vitamin D deficiency are obese people or people with dark skin tones, older adults those who get limited sun exposure and people with chronic diseases. The most common signs and symptoms of vitamin D deficiency include increased bone fractures, weakend immune system, weakened muscles, impaired healing, soft bones, hair loss, more prone to infections and tiredness.Higher levels of vitamin D in the body can lead to a condition called hypercalcemia with symptoms like nausea, headache, weight loss, damage to heart or kidneys, reduced appetite and high blood pressure (Winzenberg and Jones (2013).

Vitamin E is an antioxidant that helps the body to destroy free radicals. Free radicals are unstable atoms that may cause the formation of cancer cells. This vitamin plays an important part in preventing cancer.Vitamin E is broken down into eight different types, with the two main kinds being tocopherols and tocotrienols. Tocopherol contains the most abundant form of vitamin E. As an antioxidant, vitamin E protects the fatty tissues from free radicals that can cause cancer. Some water-soluble vitamins like B and C, help in the functions of vitamin E. In higher doses, vitamin E functions as a blood thinner. Vitamin E can be obtained from seeds, vegetable oils and nuts. Some of the best sources of vitamin E are wheat germ oil, sunflower seeds/ oil, hazelnuts and almonds.The recommended daily values are for infants aged 0–6 months: 4 milligrams (mg); infants aged 7–12 months: 5 mg;children aged 1–3 years: 6 mg;children aged 4–8 years: 7 mg; boys aged 9–13 years: 11 mg;14 years old and above: 15 mg and during lactation: 19 mg. Vitamin E deficiency is not found in healthy individuals. Those with specific illnesses that block the liver from absorbing vitamin E are at risk.Symptoms of deficiency include: trouble during walking, muscle weakness or tremors, vision issues and numbness. Also many long-term health issues can result from vitamin E deficiency that include anemia and heart disease.People who take blood thinners should be careful as overdose could increase the risk of developing cancer.

Vitamin K is present in Kale, butter, egg yolks, spinach and parsley all contain vitamin K which helps in blood clotting. The two common groups include vitamin K-1, found in plant sources, vitamin K-2, found in animal sources. There are additional man-made types of vitamin K. It can also help in reducing the risk of heart diseases, bone health reducing the buildup of calcium in the blood.The recommended dose are for infants aged 0–6 months: 2 mcg, Infants aged 7–12 months: 2.5 mcg, children aged 1-3 years: 30 mcg, children aged 4–8 years: 55 mcg, children aged 9–13 years: 60 mcg, children aged 14-18 years: 75 mcg, adult women: 90 mcg.adult men: 120 mcg.Vitamin K is not stored in an amount in the body as vitamin A or D. This can lead a person to experience a vitamin K deficiency very quickly with a greater risk of excess bleeding and reduced bone density that can lead to fractures.Synthetic vitamin K-3, however, could cause overdose if taken in large quantities.In general, vitamin K is considered safe to consume. Fat-soluble vitamins play an essential role in a person's overall health.It is important to manage the amount of each vitamin to avoid deficiency and overdose.

The particular vitamin and micronutrient in specific foods are given below

Vitamins

Vitamin B-6

Bananas	Fish (most)
Liver	Meat
Nuts and seeds	Potatoes and sweetpotatoes
Poultry	Whole-grain and fortified cereals

Vitamin B-12

Eggs	Fish and shellfish
Fortified cereals	Meat
Milk and milk products	Organ meats

Vitamin D

Egg yolk	Fortified cereals
Fortified milk	Liver
High-fat fish	

Vitamin E

Margarine	Nuts and seeds
Peanuts and peanut butter	Vegetable oils
Wheat germ	Whole-grain and fortified cereals
Folate	Dark green vegetables
Dry beans, peas, and lentils	Enriched grain products
Fortified cereals	Liver
Orange juice	Wheat germ
Yeast	

Vitamin K

Broccoli	Brussels sprouts
Cabbage	Leafy green vegetables
Mayonnaise	Soybean, canola, and olive oils

Minerals

Iodine

Iodized salt	Saltwater fish and shellfish

Magnesium

Cocoa and chocolate	Dark green vegetables (most)
Dry beans, peas, and lentils	Fish
Nuts and seeds	Peanuts and peanut butter
Whole grains	Phosphorus
Dry beans, peas, and lentils	Eggs
Fish	Meat
Milk and milk products	Nuts and seeds
Poultry	Whole grains

Zinc

Dry beans, peas, and lentils	Meat
Poultry	Seeds
Shellfish	Whole-grain and fortified cereals.

Deficiency or higher quantities of any of the nutrients can change the metabolism in the body. The deficiency of any one or more nutrients can cause malnutrition. There are many causes of malnutrition.

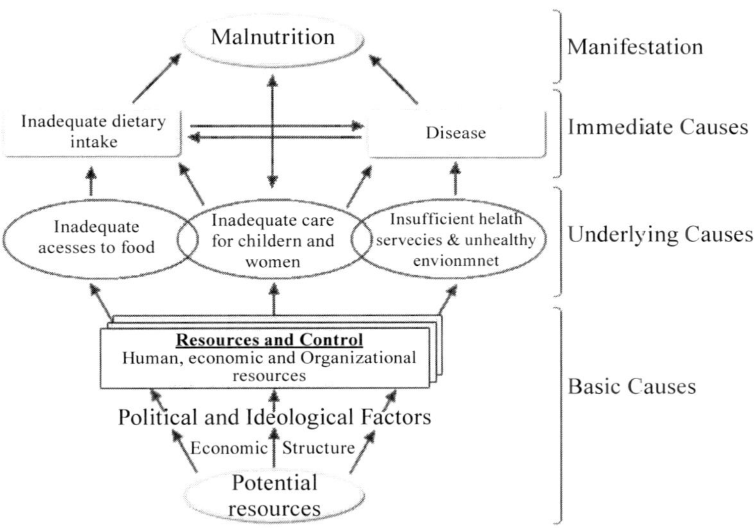

Fig. 1: Source: Causes of Malnutrition - UNICEF Framework

5

Policy Analysis for Nutritional Security Role of FAO and WHO

To look into the nutritional security around the globe, FAO and WHO, have undertaken the analysis of policies. They have covered many issues which include; World Summit on Food Security, pillars of food security and availability, access, utilization, stability, effects of food insecurity, stunting and chronic nutritional deficiencies, depression, anxiety and sleep disorders. There are many challenges to achieve food security. These include global water crisis, land degradation, climate change, agricultural diseases, food versus fuel, politics, food sovereignty, food wastes, risks to food security, population growth, fossil fuel dependence, homogeneity in the global food supply, price setting, land use change, global catastrophic risks, agricultural subsidies and role towards children.

According to Holleman and Conti (2019) It has been found that women usually take the responsibility for food preparation and childcare within the family and are more likely to spend their income on food and their children's needs.) They play an important role in food production, processing, distribution and marketing often working as an unpaid family worker. In the agriculture sector they are involved in subsistence farming and represent around 43% of the agricultural labor force in developing countries. This varies from 20% in Latin America to 50% in Eastern and Southeastern Asia and Sub-Saharan Africa. However, women face discrimination in access to land, credit, technologies, finance and other services. According to many studies it appears that if the women would have the same access to productive resources as men, then the women could boost the yields by 20–30%, raising the overall agricultural output in developing countries by 2.5 to 4%. While these are rough estimates, there would be a significant benefit of closing the gender gap on agricultural productivity. de Onis and Branca (2016) have discussed a global perspective for childhood stunting.

Black et al., (2013) have looked into the reasons for maternal and childhood undernutrition. The gendered aspects of food security are visible along the four pillars of food security: availability, access, utilization and stability, as defined by the Food and Agriculture Organization (Young, 2013). The Quarterly Journal of Economics has discussed the inequality, the urban-rural gap and migration. The recent statistics (Young, 2013) shows that the number of people affected by hunger are extremely high with girls and women in large numbers (Young, 2013). Many international institutions and governments are working on it. Food security is concerned with equality, rights and social justice. Factors like capitalism, exploration of indigenous lands contribute to food insecurity among the minorities and the people who are the most oppressed in various countries (women being one of these oppressed groups. It is important to control the lower-class via the food market. According to Young (2013) this could happen only when the government can lower the prices. He has used the population and product consumption data from the demographic and health surveys.He was able to make comparable measures of inequality and migration for 65 countries, including some of the poorest countries in the world. It was found that the urban-rural gap accounts for 40% of mean country inequality. Individuals who were raised in rural areas and moved to urban areas were found to earn much higher income than nonmigrant rural permanent residents and vice versa. The population sorts itself geographically on the basis of its human capital and skill. This has explained the urban-rural gap in living standards. In order to achieve food security and to end hunger global partnership is needed. It was in April 2012, that the Food Assistance Convention was signed which was the world's first legal binding international agreement on food aid. Later recommendations were made for the efforts to combat hunger and malnutrition on priority with maximum effectiveness of aid spending.The Commission on Social Determinants of Health (CSDH, 2008) (WHO, 2008) studied the health equity on the social determinants of health.The role of social determinants to tackle health objectives during economic crisis is well discussed (Marmot et al., 2013). Recently, the major global policy to reduce hunger and poverty are the Sustainable Development Goals. Particularly the goal 2 of Zero Hunger sets globally agreed targets to end hunger, achieve food security, improved nutrition and promote sustainable agriculture by 2030 (Popay et al., 2008). Kristjansson et al., (2006) made a survey on school feeding for improving health in school children.

According to the UN, Sustainable development Goal 2; a profound change is needed in the global food and agriculture system to nourish the 821 million people who are hungry today. Apart from this, there are additional 2 billion people which are expected to be undernourished by 2050. Investments in agriculture are crucial to increasing the capacity for agricultural productivity and sustainable food production systems are necessary to help alleviate the perils of hunger.A number of organizations have formed initiatives with the goal to achieve this by 2025.In 2013 Caritas International started a Caritas-wide initiative aimed at ending systemic hunger by 2025. Food for all campaign focuses on awareness, improving the effect of Caritas programs and advocating the implementation of the right to food for all. (Commission on Social Determinants of Health (CSDH), 2008). The partnership compact 2025, led by IFPRI with the involvement of UN organisations, NGOs and private foundations develops and disseminates evidence-based advice to politicians and other decision-makers aimed at ending hunger and undernutrition in the coming 10 years, by 2025. A conceptual framework for national agricultural, rural development, and food security strategies and policies has been prepared (Stamoulis and Zezza. 2003).

In June 2015, the European Union and the Bill & Melinda Gates Foundation had launched a partnership to combat undernutrition, especially, in children. In another program the Food and Agriculture Organization of the UN has created a partnership that will act through the African Union's CAADP framework aiming to end hunger in Africa by 2025. It includes different interventions like support for improved food production, strengthening of social protection and integration of the Right to Food into national legislation (FAO/UN, 2018 African Union High-Level Meeting, Addis Ababa, Ethiopia).Improving agricultural productivity to benefit the rural poor needs to be looked into and also to reduce large-scale food stockpiling. It is important to have Agricultural insurances.These policy guidance notes have been produced in the frame of the strategic partnership between the Food and Agriculture Organization of the United Nations (FAO) and the Directorate for International Cooperation and Development of the European Commission to boost food and nutrition security, sustainable agriculture and resilience. This policy guidance note was prepared by Dubravka Bojic and Klaus Urban of FAO, with contributions from Jean Balié, Lalita Bhattacharjee, Michael Clark, Frédéric Dévé, Ciro Fiorillo, Juan Carlos Garcia Cebolla, Yasaman Matinroshan, Jamie Morrison and Esther Wiegers. This information product do not imply the expression of any opinion on the part of FAO.FAO information products are available on the FAO website (www.fao.org/publications) and can be purchased through publications(sales@fao.org). This policy guidance note is part of a series that the Food and Agriculture Organization of the United Nations (FAO), the Directorate for International Cooperation and Development (DEVCO) of the European Commission and partners are producing to support the policy makers in order to address the food security and nutrition situation in their country. Guidance is provided with each note to sharpen the focus of sector policies so that sustainable food security and nutrition outcomes can be achieved.The note has introduced the challenges of political economy to the persons engaged in food and agriculture. It provides practical knowledge as to how political economy analysis (PEA) can be undertaken and to address the political challenges and dynamics around food and agriculture. The global food crisis of 2008 and after, and the challenges that many governments faced to implement effective national policy has responded to the food price volatility, re-focused attention on the political economy of policy-making and implementation for food security and nutrition (FSN). Many times technically sound policies may not produce good results. The lack of progress in FSN practices is mainly due to insufficient role of politics, interests and participation of the institutions. In a FAO report (2009) of FSN in five countries showed that a good understanding of local dynamics, their influence on the policy process and the various interests play a major role and influences the organization's ability in engaging into sectoral policy processes and mainstream FSN (Balié, 2009). However, PEA can be useful in the identification of potential entry points which will support the efforts of reformation. It explains the public about PEA and how is it helpful in value addition for the support work of the policies which will improve the food security and will provide better nutrition at the country level. UK Department for International Development (White Paper on International Development, 2006) asserts that it is the politics which determines the use of re sources and how policies are made. It also says that politics determines the beneficiaries (DFID, 2006). Important efforts have been made to understand how political processes and institutions affect development trajectories and influences the policy choices (Adam and Dercon, 2009). The literature on political economy of development explains the qualitative and quantitative approaches, frameworks, tools and case studies prepared by development agencies. It relates to the social, economic, cultural and political factors that structures, sustains and transforms the constellations of public and private institutions and their interests over time. The various political economy have different analytical tools and methods but they have some commonalities. One common theme is their identification of different layers of analysis. Day-to-day politics is affected by shaping the policy agenda like the history of the country under discussion, its natural resource endowment and the power distribution across social groups (Hout, 2015). Recently, there have been many approaches which point to the relevance of ideologies and policy ideas (Review of the Role of Political Economy with respect to Agricultural Policymaking in Africa (David, 2018; Booth, 2015; Booth, 2012, 2014; Booth and Cammack, 2013; Booth et al., 2015 and Bordes et al., 2008). The classical political economy emerged in the eighteenth century during industrialization and when the capitalist economic system in Europe were being established. PEA has been worked out in depth by the DFID, Swedish International Development Agency, German Agency for Technical Cooperation (GTZ) and the World Bank. During the globalization with an increased number of stakeholders and fast changes in economic, social and political scenes are creating a very complex world. For the realization of FSN it is important to be pursued in a politically informed manner. To achieve that, there should be a practically oriented PEA which can explain political processes and could bargain between public and private sectors, to understand their relative power and the institutional context within which they operate.

Non imposing approach to public can facilitate building trust and motivate them towards success. Continuous dialogue with the concerned stakeholders on all sides of the policy issue are the key.Food security continues to be a growing problem with growing population. For the formulation of effective policies a thorough understanding of the empirical data and the application of appropriate measurement and analysis are required (Babu et al. 2014). The Food and Agriculture Organization of the United Nations (FAO) and the Directorate for International Cooperation and Development of the European Commission have jointly worked to boost food and nutrition security, sustainable agriculture and resilience (www.fao.org/contact-us/licence-request or addressed to copyright@fao.org.FAO information products are available on the FAO website (www.fao.org/publications) and can be purchased through publications(sales@fao.org. ©FAO, 2017). In 2000, leaders worldwide adopted the Millennium Declaration and the Millennium Development Goals (MDGs). At the end of 2015, the prevalence of hunger had globally declined from 15 percent in 2000-2002 to 11 percent in 2014-2016. As the MDG's expired, world leaders adopted the 2030 agenda for, Sustainable Development and the 17 Sustainable Development Goals (SDGs).This is a very ambitious goal which is aimed to end hunger, achieve food security with improved nutrition. It would promote sustainable agriculture by 2030 (SDG 2). The access to safe, nutritious and sufficient food at all times of the year, requires sustainable food production systems and resilient agricultural practices, equal access of men and women to land, technology and markets, and investments in infrastructure and technology to boost agricultural productivity. The achievement of this goal is closely linked to the achievement of SDG 1, which aims to end poverty in all its forms everywhere. To achieve SDG 1, there is a need to ensure social protection for the poor and vulnerable, increase access to basic services and support people who are negatively affected by climate change and other economic, social and environmental shocks and disasters. The MDG experience shows that under a scenario where current trends continue, the twin goals of ending hunger and poverty by 2030 will not be achieved. To achieve these goals, greater focus is needed in policies and programmes which would have an explicit and measurable food security and nutrition objectives. These should be combined with a significant increase in resource allocation, investment, intensified dialogue and collaboration across different sectors, stakeholders and development partners that are relevant to these common goals.

In this context the Food and Agriculture Organization of the United Nations (FAO) and the European Union (EU) supports the governments jointly along with their development partners. They help in creating a policy and institutional environment which is conducive to achieve SDG 1 and 2 with the implementation of the "Food and Nutrition Security Impact, Resilience, Sustainability and Transformation" (FIRST) programme. To this end, the first programme supports a variety of activities, which covers reviewing the existing national policy and institutional frameworks for food security, nutrition and sustainable agriculture, identifying bottlenecks and opportunities for improving policy and programme impact, advocating for and developing human and organizational capacities for food security, nutrition and sustainable agriculture, facilitating evidence-based inclusive policy dialogue and stakeholder coordination, promoting more investment and improving resource allocation in line with the priorities spelled out in the national policies. Under policy guidance series the guidance notes will support FAO and EU policy officers and their partners in the efforts to promote inclusive and evidence based stakeholder dialogue and in the development of capacities to improve the enabling environment for food security, nutrition and sustainable agriculture.

Leonard Mizzi Karel Callens (Head of Unit Rural Development, Deputy Leader Food Security, Nutrition Strategic Programme to eradicate hunger) has developed a set of policy guidance notes to improve the coherence and the contribution of national policy frameworks towards the achievement of SDGs 1 and 2. Each of the guidance notes analyses a specific policy domain and outlines the policy options for improving the impact on food security, nutrition and sustainable agriculture. These have emphasis on the food security, nutrition and sustainable agriculture. The implications on the short, medium and long terms of various policy options, trade-offs between short-term and long-term objectives, between the needs and demands of different stakeholder groups and domestic priorities with international obligations. Issues pertaining to governance and stakeholder dialogue around specific policy issues and processes need to be seen, issues of political economy in terms of the actions and interactions among various stakeholders, coalitions and networks that are involved in specific policy processes, each with their own agendas, interests, values, beliefs, relationships, political power, knowledge and capacities.

USD 1 invested in a bundle of evidence-based interventions for nutrition generates average returns of USD 15 (Hoddinott et al., 2012). Such investments include nutrition-specific interventions with proven effectiveness and is related to behaviour change, provision of micronutrients and complementary/ supplementary and therapeutic feeding interventions. When the nutrition-specific interventions are implemented effectively then it will reduce stunting and micronutrient deficiencies to around 20 percent (Bhutta, et al., 2009; Global Panel, 2016). Therefore, to meet the various nutrition targets policy makers must combine nutrition-specific interventions with effective actions in other key sectors that address the basic causes. World Health Assembly implemented the plan on the nutrition of maternal, infant and young children with a set of six global nutrition targets to be completed by 2025. This will help in the reduction of 40 percent of stunted children under five, a 50 percent reduction in anaemia among women of reproductive age, a 30 percent reduction in low birth weight, ensure that there is no increase in childhood overweight, increase the rate of exclusive breastfeeding in the first six months upto 50 percent, reduce and maintain childhood wasting to less than 5 percent.Causes of hunger, food insecurity and malnutrition needs to be addressed by the policy makers. The most important issues are agriculture, fisheries, aquaculture, livestock, forestry and land tenure. The EU has committed to support partner countries in reducing stunting by at least seven million people by 2025, corresponding to 10 percent of the World Health Assembly target. Taking this objective the EU has prepared the nutrition policy framework on “Enhancing Maternal and Child Nutrition in external assistance: an EU policy framework”.

Policy makers need to address the impact of these trends. They need to look into the challenges it will have on the food systems and consumers. Ultimately, it can give many good opportunities for improvement. In terms of their main impact on food security and nutrition, the emerging trends and challenges fall into three broad categories, namely: (i) changes in demographic structures and pressures from population growth, especially in resource-constrained countries that have high population growth rates (ii) urbanization, changing lifestyles and consumption patterns (iii) climate change and natural resource competition. With the growing world population, many demographic changes have taken place and therefore, accordingly, the food systems need to be in pace with the increasing demand for food. This has brought in a need for changing dietary patterns which must provide incentives for (i) research, development and transfer of technologies towards sustainable increase in production (ii) investments to make food supply chains more efficient for consumers’ needs in ways that increase people’s access, especially for the poor and vulnerable people, food that is acceptable, affordable, nutritious, safe and healthy. Urbanization, changing lifestyles and consumption patterns have changed a lot depending on the type of food consumption and physical activity. Sedentary lifestyles, consumption of highly processed convenience foods high in dietary energy and poor nutrition make a person unhealthy. Looking into the factors in the policy framework that affects consumer behaviour and lifestyle plays a role in preventing overweight, obesity and the resulting soaring morbidity. It increases the risk of food-borne diseases. There is a need to enhance the domestic food infrastructures and systems to protect consumers. The public sector policies and regulations must ensure that private sector policies do not harm food security and must give good nutrition and health. Climate change can increase the limitation of natural resources in rural areas which would affect their livelihoods and nutrition.Food security, can be achieved only when all people have physical and economic access to sufficient, safe and nutritious food at all times and at all levels in the food chain. This will help towards an active and healthy life references (FAO, 1996).This includes availability, access, stability and utilization of food. The supply of food and a country's ability to provide enough nutritious food to meet the needs/demands of the population is determined by the level of food production or imports, stock levels and access to the markets.Good childcare and hygiene practices along with diversity of the diet influences the nutrient intake by the individuals. Under chronic food insecurity persons are not able to meet the minimum food requirements over a long period of time, primarily as a result of poverty. However, under transitory food insecurity there is a sudden drop in the ability to either produce or access enough food to maintain a good nutritional status. This could happen due to climatic shocks, economic collapse or violent conflict.

Nutrition is known as the intake of food and the interplay of biological, social, and economic processes that influences the growth, function and repair of the body (FAO, 2013). Food security is an important goal as a precondition for nutrition security. The relationship involves proper nutrition.The Road Map to improve nutrition has been worked out by the American Dietetic Association (2007). According to them it can be achieved with there is a secure access to a nutritious diet. It should be prepared under a sanitary environment with adequate health services. However, food security can be obtained without looking into health or sanitation. Many different forms of malnutrition are undernutrition; the outcome of insufficient food intake to meet dietary energy requirements and/or poor absorption or biological use of nutrients as a result of repeated infectious disease. These include chronic (resulting from long-term inadequate intake and/or repeated infections) or acute (resulting from a rapid deterioration of nutritional status due to sudden food deprivation and/or bout of illness). There can be a specific form of undernutrition called as micronutrient deficiencies which occurs by the lack of vitamins, minerals and/or trace elements.Although these are required in small amounts but are found to be essential for the proper functioning, growth and metabolism of a living organism. Overweight and obesity is kown as a condition of body weight that is above the weight given by the doctors for the person's height which happens due to the excessive accumulation of fat.

Food insecurity can be at the level of household (in terms of food availability, access, utilization and stability), inadequate care and poor availability of clean water, sanitation and health services. It also includes environmental factors, like climate change. They include factors like governance and institutional capacities, gender, social mechanisms, access to education, presence of infrastructure, trade policies and systems. Conceptual framework of malnutrition emphasizes the importance of basic causes like the linkages between political and social factors. According to WHO a healthy diet helps to protect against malnutrition and non-communicable diseases (NCDs) (Lachat et al., 2013). More specifically, a healthy diet is characterized by an adequate consumption of health-promoting foods, food groups and nutrients and moderation in consumption of foods, food groups and nutrients which are linked with poor health outcomes (Guenther et al., 2008a). The scientific evidences support that dietary risks decrease if fruits, vegetables, whole grains, nuts and seeds, milk, legumes and pulses, sea foods and polyunsaturated fatty acids are eaten in adequate quantities. Products like red and processed meat, sugar, trans-fatty acids and salt should be eaten in small amounts. Diet is the major risk factor which contributes for deaths (Forouzanfar et al., 2015). The food availability is dependent on the food price and convenience to prepare them (Herforth and Ahmed, 2015). Policies for better food security and nutrition includes the storage and processing of food at household, community and commercial levels, related infrastructures, farmers, micro-processors, small and medium enterprises, large national and multinational businesses, regulations of these activities, public and private investments and strategies for agroindustrial development. Food trade depends on the local, domestic, regional and international levels, post harvest cold chain during transportation and at the market place, trade regulations and agreements. The post harvest chain includes processing, logistics, wholesale, retailing and fast-food catering (Reardon et al., 2003 and Reardon et al., 2009). Changes also take place due to taxes, subsidies and regulation (Pinstrupp-Andersen and Watson, 2011). This implies that there is a strong potential for public policy to improve the nutrition outcomes of the food system and to make it more nutrition sensitive.

The term policy comes from the Middle English word policie, meaning art of government or civil organization. A standard definition of policy is: "a plan or course of action, as of a government, political party or business, intended to influence and determine decisions, actions and other matters" (West's Dictionary of American Law). More specifically, a public policy can be defined as "an intervention of a public authority aimed at affecting the normal course of events" (Birkland, 2010). Public policies are implemented through the adoption of policy measures which include; taxes, subsidies and regulations. It is important that the policy measures stimulate the adoption of new technologies.They should help in the export of products. It is also very vital to preserve assets and resources by using sustainable agricultural techniques. These include rotation of crops, soil fertility, vaccinations, waste-water treatment, appropriate disposal of hazardous wastes, not using hazardous pesticides, have selected imports or exports and not having excessive energy use. The policy must be well taken at all levels of governance, from municipal to intergovernmental. It can be revealed through legislative texts, policy statements, documents, administrative practices like sector plans, programmes, budgets and changes in high level rules of government agencies (Schneider and Ingram, 1997). Policies are formulated and implemented through policy processes. A policy process is the way in which policy changes are planned, designed, implemented and evaluated. Given the fact that public policies involve many persons who are interlinked through socio-economic or political relationships, a policy process is multi-faceted, complex and reflective of the social and political contexts in which it takes place (Forouzanfar et al., 2015). Collective policy action needs to be taken among key sectors like agriculture, fisheries, forestry, health, education and in policy domains like social welfare, land tenure, gender, women's empowerment and climate change towards enhancing food security and nutrition. The sector/ policy domain-specific policy instruments address different aspects of the underlying structural causes based on their mandate, while trade-offs between the specific policy objectives and food security and nutrition objectives are recognized and resolved and synergies are effectively exploited. By definition, a cross-sectoral policy framework for food security and nutrition cuts across different sectors. In principle, these are virtually all sectors affecting the structural causes of food insecurity and malnutrition in a particular context. However, each of these sectors pursues its own goals, which may not always be in harmony with food security and nutrition objectives. Due to lack of understanding and weak cross-sectoral coordination, food security and nutrition concerns are not always sufficiently factored into specific-sector policies and decision-making. At the same time, sectoral concerns do not always adequately feature on the food security and nutrition agenda. The result is that proposed policies and programmes do not include alternative instruments and pathways, which would maximize their impact on food security and nutrition given the objectives of these policies and programmes. Therefore, promoting a coherent approach calls for sharpening the policy focus of the relevant sector policies in terms of their specific contributions to food security and nutrition. In other words, these sector policies are not only appraised and evaluated from the perspective of their specific policy objectives, but also with regard to objectives related to food security and nutrition. In practice, implementing a coherent and cross-sectoral framework for food security and nutrition is complex and challenged by serious gaps in detailed knowledge of the actual and potential impacts of specific policy measures on categories of food insecure and malnourished populations. Further, capacities to recognize synergies and resolve trade-offs among objectives of various sector policy instruments and food security and nutrition objectives are limited, especially given the overall association of food security with food production, agriculture and nutrition with health. This policy uncertainty coupled with conflicting political agenda across sectors puts a limit on how much can be done to promote coherence, even in one sector with various sub-sectors. Thus, in practice one is limited to ensure coherence among the key sectors that influence the food security and nutrition outcomes. An example of a country that adopted a coherent and crosssectoral policy framework, reflects, the contributions of selected sectors to food security and nutrition.The core of any strategy for greater policy coherence is to agree on common policy objectives that addresses both sectoral needs of food security and nutrition-related challenges in a coordinated and consistent manner (Hawkes, 2015). In this respect, policy coherence requires the engagement of a broad range of policy users. It concerns in identifying and overcoming the tensions and conflicts within and across different sectors or sub-sectors that might undermine food security and nutrition outcomes and facilitate synergies across policy measures. It implies that decision-makers and other stakeholders have a common understanding of food security and nutrition problems and solutions in the country which will recognize and understand the interactions between different sectors and their specific impacts on food security and nutrition.

The policy coalitions are interrelated and therefore the individuals and networks are continuously engaged and influence each other. The policy, if changes, tends to emerge in an unpredictable manner. Any economic crisis can force the policy-makers to change it completely. However, influencing policies in the agriculture and related sectors brings its own particular challenges. The agriculture sector, in many developing countries, depends on the public concerns (Wiggins et al., 2013). In most low-income countries it employs much of the workforce and contributes significantly to the gross domestic product and agricultural practices and land use are often the subject of important environmental concerns.Consequently, a wide range of objectives are commonly invested in agriculture and rural development: economic growth and export earnings, employment and equality for all. The Fome Zero (Zero Hunger) policy framework in Brazil was introduced to eliminate hunger and to improve the livelihoods of the poor. The Brazilian government placed food security and nutrition as a strategic objective of public policies in order to promote a coherent and coordinated approach and integrate support programmes focusing on agriculture, nutrition, health, education and poverty reduction. The Zero hunger policy mainly focuses on two areas: (i) improving access to food among low-income populations and (ii) strengthening family farming, the main source of food supply to the domestic market. For this the Organic Law on Food and Nutritional Security was implimented by the government to ensure the right to adequate food. A law was brought forward on Family Agriculture in 2006. A new Ministry of Social Development and Fight against Hunger was set up for better coordination among all the ministries for the national food and nutritional security policy bringing in more commitment (de Schutter, 2014). This is important to end hunger, food insecurity and malnutrition and to manage conflicting policy agendas. It is important to have evidence based analysis and relevant data related to food security and nutrition implications of policy actions. This will help in decision making and to translate political commitments into practice. The analysts must identify, describe, analyse and assess the impacts of policy options on food security and nutrition in the short and long terms. It is important to highlight the trade-offs and synergise the sector policy aims, food security and nutrition. They must propose adjustments in the policy instruments and suggest measures which could improve food security and nutrition outcomes. This would communicate relevant analytical findings to the decision makers and other stake holders who are involved in decision making processes. They could then have policies and policy dialogue among the stakeholders which will reduce the scope of ideology. Policy processes are organized to ensure multi-stakeholder dialogue. Policy dialogue among analysts, decision-makers and other stakeholders must reach a common understanding of the impacts, synergies and tradeoffs to identify the barriers in policy change. Coordination mechanisms will take into account the food security and nutrition concerns. They will facilitate synergies and will resolve potential policy conflicts. The sector policies for better food security and nutrition should be strengthened. According to Hummelbrunner and Jones (2013) the policies are needed on gender equality, food and nutrition security, environmental conservation and regional equity. It is found that the information on processes and positions of key factors is incomplete and therefore the objectives and approaches need to be changed. The engagement work must be carried out in a network format with a specific role of any organization.An inside-track approach can be differentiated from that of an outside-track approach (Start and Hovland, 2004). Inside track involves methods using direct interactions with decision makers and the partners. They include participation in negotiations, meetings, direct communications with government ministers or informal discussions with collaborators. Outside-track, targets a larger number of individuals on an issue, through public messaging and campaigning. Use of public meetings and speeches to communicate the rationale for a proposed reform can be helpful.

They may use television and radio to raise public awareness of an issue. Policies can be shaped with the involvement of a proactive politician, bureaucrat, an economist and social interest group along with a citizen's campaign that gains traction in civil society and the media. Most of the times it is the civil servant who either makes or stalls a policy. In order to work with public agencies it is better to target and build relationships with second or third tier officials as they are helpful in re-structuring and social-psychological interventions in education (Yeager and Walton, 2011). The FAO-ODI handbook on policy engagement and influence is a modification of the ODI RAPID Outcome Mapping Approach (ROMA) and comprises of a suite of tools. These can be used at any stage of policy engagement. It also shows a way to learn a problem, understand the type of impact, set realistic objectives for influence, develop a plan to achieve the objectives. It will be better to consider one's policy engagement process as a compass navigating through climate change (Smith et al., 2013; Young and FAO, 2015). In order to enhance the Nutritional Impact of Agriculture Investment Programmes, a checklist and guidance for programme formulation has been given by Forouzanfar et al., (2015).Guenther et al., (2008 a) have discussed on the urgency of policy action to be taken. Hawkes (2015) have looked into the development of the healthy eating index. Herforth (2016) has discussed on the enhancment between the trade policy and nutrition. In the UN Standing Committee on Nutrition (Herforth and Ahmed, 2015) the need for assessing the impact of food systems policies on food environments and diets was discussed with emphasis on the food environment, its effects on dietary consumption and potential for measurement within agriculture nutrition interventions. FAO, (Balié, 2009) have discussed the strategies on food and nutrition security. A methodological guide for policy intelligence and preparedness work in pilot countries had been drafted in Rome by FAO (Bhutta, et al. 2009). Birkland (2010) had studied an evidence-based interventions for the improvement of maternal and child nutrition. Herforth, 2015 and Dubé, et al., 2014) have studied the global strategic framework for food security and nutrition. European Union (EU, 2013) looked into the coherence policy which will provide human and economic development. European Commission (FAO, 1996) prepared an EU Policy Framework to enhance maternal and child nutrition in external assistance. FAO, (2013) gave a Rome Declaration on World Food Security and World Food Summit Plan of Action in 1996. FAO/WHO, (2014) have given the conference outcome document which explains the framework for action on nutrition.

The book, Global food policy report international food policy research institute (ifpri, 2018) has provided the global food policy report in an excellent form.It reviews the major food policy issues, developments and decisions of 2017 and highlights the various challenges and opportunities for 2018 at the global and regional levels. The report looks at the impacts of greater global integration and movement of goods, investment, people, knowledge and the threat of antiglobalization pressures.

Under the Guidelines by FAO and EU, 2017 (Strengthening Sector Policies for Better Food Security and Nutrition Results: Political Economy Analysis. Policy Guidance Note 8 by FAO and EU. 2017, Publisher: FAO, European Union; http://www.fao.org/3/a-i7212e.pdf) there is an understanding of the domestic political economy to promote policies which are adapted towards the realities. The challenges that many governments have been facing in response to food price volatility during the 2008 global food crisis has brought the attention to the political economy and implementation of effective national policy responses on food security and nutrition. Understanding the political economy behind public policy-making and implementation is crucial in order to enhance the effectiveness of policy support to the countries. This increases the chance of nationally-led reforms that results in better food security and nutrition outcomes. The political economy analysis was intended to support policy practitioners and other stakeholders to work in politically informed way. According to the report it will be coherent set of policies across relevant sectors. It has also demonstrated the political economy analysis in relation to policy support work. It shows ways towards key political economy challenges in food security and nutrition and also on the existing frameworks and analytical tools. The efforts undertaken to eradicate hunger in recent decades in Latin America have allowed for achieving the objectives established by the World Food Summit and in the Millennium Development Goals. The region has obtained vast technical and political experience to combat hunger and malnutrition. These could be achieved due to the political commitment and the social involvement. Within each country many measures, policies and programmes have been established which will promote food and nutrition security (FNS) with a multisectoral approach, seeking the participation of a variety of social partners. The countries with the highest percentage reductions among undernourished populations between the 1990-1992 and 2014-2015 biennia are Nicaragua (54.4% to 16.6%), Panama (26.4% to 9.5%) and Mexico (6.9% to 4.2%) in Central America, along with Peru (31.6% to 7.5%), Bolivia (38% to 15.9%) and Brazil (14.8% to 1.5%) in South America (FAO, 2015). These efforts have also reduced child malnutrition in the region to 12 percent. However, Latin America faces the challenge of malnutrition which can be seen in more than 26 million people with undernourishment, children (6 million) with chronic malnutrition and another 3.6 million people with issues of overweight (FAO, 2015). Ending hunger, achieving food security, improving nutrition and promoting sustainable agriculture are the main objectives framed under the Sustainable Development Goals (SDGs) which were approved by the international community in 2015 and now comprise the new global agenda. This will help to eradicate poverty, protect the planet and ensure prosperity over the next 15 years. For Latin America, the SDG's support the continuity of the actions currently underway. It is important to understand the main constraints that have delayed the attainment of food and nutrition security and the human right for adequate food. Undernourishment is considered to be an insufficient quantities of food intake.Some of the regional efforts in recent years have addressed the establishment of legal and programmatic frameworks through them. The creation of a set of entities to promote FNS, with particular commitment at the national and local levels. These have been key contributors to food and nutrition security within the priorities for action in public and private institutions, academia, cooperation agencies and other social organizations. Similar policy-making bodies have been established at different territorial levels. A weak political culture provides the discomfort which leads to lower the decision making. In the area of FNS there has been a lot of progress made which is a clear evidence of the importance of having specific legal and programmatic frameworks. Sometimes, the constraints faced need to be solved. It requires both public, private and the civil society to maintain consistent, coordinated and effective interventions. It is vital to analyze the factors and mechanisms for the cooperation and joint action between the various sectors that are linked to FNS. It is also necessary to understand the behavior of the people who are involved and the dynamics that exists between them towards achieving zero hunger. The FNS document assists in giving an understanding into the governance of food and nutrition security and begins its analysis and provides recommendations for improvement. Many researchers have proposed the methodological approaches to explain the processes of governance, identification of the constraints observed and seeking potential solutions.From the Policy Assistance and Governance Unit of FAO, the following methodological steps are proposed (FAO, 2011): Define the problem, taking into account that the definitions are not neutral. This task is critical in the process in order to avoid individual influences and subjectivities. Understand the formal or informal rules which govern and have an impact on societal actions. Characterize the actors, their influential capacity and how they interact with one another. Identify the convergence of the entities, how their spheres of action are shaped, and how they operate. Understand the intervention processes and behavior to generate and improve the system. Latin America region has a Framework Law on the Right to Food, Food Security and Sovereignty which has been adopted by the Latin American Parliament (Parlatino). At the national level, many countries possess national laws on FNS or food sovereignty which include Guatemala, Brazil, Ecuador, Nicaragua, Venezuela, Honduras and Dominican Republic. Mexico City has also approved a law on FNS. In Argentina, Peru, Bolivia,

Colombia, Paraguay, El Salvador, Costa Rica, Panama, Haiti and Mexico, draft laws on FNS are currently under consideration. Nine countries in the region, mention the exercise of the Human Right to Adequate Food (HRAF) for their citizens (CELAC, 2017). The promotion of these legal frameworks has involved both political and social persons. These regions have advanced in the implementation of legal frameworks for FNS. Certain countries have made structural or functional adjustments to new FNS. Apart from the difficulties, the legal and regulatory frameworks have established mechanisms of governance that are adapted to national and local realities and seek to be the driving force for social participation and involvement.

FAO learning LAB (2016) mapped and analyzed the policies for FNS. In the Second International Conference on nutrition (ICN2) which was held in November 2014, all the world leaders suggested for a greater policy coherence to ensure that policies, strategies and activities of each relevant sector contributes and reinforces the work of other sectors in reducing malnutrition (Rome Declaration on Nutrition, 2014). In September 2015, they called for enhancing "policy coherence for sustainable development" for the implementation of the new Agenda of 2030 and achievement of the Sustainable Development Goals. It was in 2016 that the Decade for Action on Nutrition was announced and the Committee on World Food Security (CFS) was made to promote policy convergence of multiple sectors as well as policy coherence for ensuring the contribution of food systems towards improved nutrition. Due to the gaps in knowledge and research it has limited the policy decision-making. For proper monitoring it is important to know which policies are in place and influence FSN, if they are actually being implemented or not, their impact evaluation and what are the objectives for the learning lab. It is important to sensitize the importance of policy coherence to promote nutrition sensitive food systems. Bringing attention to current efforts in terms of policy mapping and analysis is vital. In order to inform policy decision, it is important to identify priority research areas which are related to policy coherence and for the mapping and analysis for nutrition-sensitive food systems.Formulation to support linkages between research and policy are required which will help to ensure research effectively in the policy-making on food security and nutrition. According to Dubé et al, (2014) if there is consistency, comprehensiveness giving rise to harmonious and compatible outcomes across policy areas and sectors then the goals are called as "Policy coherence". Hawkes, (2015) has explained that the core of any policy coherence is to agree on common policy objectives aiming to address both sectoral policies, food security and nutrition related challenges in a coordinated and consistent manner.

Food systems encompass all the people, institutions and processes by which agricultural products are produced, processed and brought to consumers (FAO SOFA, 2013). Activities are driven by the behavior of farmers, processors, traders, consumers, policy makers and officials in government and non-governmental organizations.They have differential access to power and resources and respond to opportunities, risks and constraints imposed by the public policy. Among the main functions of food system are availability, affordability, convenience and desirability of various foods. Through the food environment, the food system influences the diets and nutritional status of the consumers'. Policies are in support of agro-processing and retail sector that can increase availability, access and affordability of energy dense and nutrient foods.In order to build coherent and cross-sectoral policies for FNS the focus must be kept on FNS objectives of a specific policy which would provide social protection to agriculture, livestock and education. This will be a key towards poverty eradication, inequality reduction and rural development. In order to empower women and make them recipient of social protection benefits strategies that enables the households to diversify their diets and livelihoods, strengthen the linkages to health and sanitation services, integration of nutrition education and scale up of safety nets in times of need should be promoted. http://socialprotection.org/connect/communities/working-group-food-security-and-nutrition-specific-inter-agency-social. In a Protection to Production (PtoP) project (http://www.fao.org/economic/ptop/home/about/en/) the findings show that coherent agricultural and social protection interventions can tackle poverty and hunger. Interventions made at the agricultural levels can address constraints in the agricultural sector whereas complementary social protection interventions can help to enhance the ability to plan expenditure. Improving production, strengthening risk management capacities and increasing food consumption and dietary diversity would help at the household levels. From Protection to Production (PtoP) project is a multi-country impact evaluation of cash transfers in sub-Saharan Africa from collaboration between FAO, UNICEF and the governments of Ethiopia, Ghana, Kenya, Lesotho, Malawi, Zambia and Zimbabwe. According to Hawkes, (2015) better analysis of the coherence between sectoral policies and food security and nutrition enables a common understanding of the opportunities and risks. An approach of solution-problem tree, can help in bringing different stakeholders together around common objectives and to establish a shared narrative.

6

Importance of Traditional Spices in Foods for Nutritional Security

Indian Spices: These are dried, fragrant, aromatic or pungent vegetables or plant substances. They can be whole, broken or in ground forms. Primary function in food is seasoning rather than nutrition and which contributes to the flavour and piquancy in foods and beverages. These have been used as food additives widely since ancient times [http://www.imedpub.com/articles/herbal-medicines-possible-risks-andbenefits.pdf,https://www.ncbinlm.nih.gov/pmc/articles/PMC3170500/pdf/nihms307509.pdf). They also exert digestive stimulant action of food enhancing the taste. A few medicinal properties of spices are as being a tonic, carminative, stomachic, diuretic and antispasmodic (Aboaba et al., 2011; Achi, 2006 and Adefegha et al., 2012). This has led for pharmacological applications in the indigenous systems of medicine as digestive stimulants and to relieve digestive disorders and makes the food more palatable. In many countries around the world these are grown and reared for the purpose of trade and distribution. In the ancient times the spice merchants have been the most popular traders. India is well known for spices with a wide variety of recipes that are laced with them.

Spices have been some of the most valuable items of trade in the ancient and medieval world. The anti-proliferative, anti-hypercholesterolemic, anti-diabetic and anti-inflammatory effects of spices have a lot of importance. In recent times, the key health concern of mankind is mainly on diabetes, cardio-vascular diseases, arthritis and cancer. The spices and their active compounds can be used as ameliorative and preventive agents for many health disorders. Spices are rich in antioxidants and are also a potent inhibitor of tissue damage and inflammation caused by high levels of blood sugar and circulating lipids. As these have very low calorie content and are not expensive, they have the reliability of providing antioxidants and other potential bioactive compounds into the diet. Kleijnen et al., (1989) have discussed the effect of garlic and onions on the cardiovascular systems. Fresh garlic and onions have a beneficial effect on cholesterol levels, fibrinolytic activity and platelet aggregation. For fresh garlic, the claims have been confirmed at high dosages whereas for onions contradictory results have been reported. Dietary spices influence many metabolic pathways in the body like gastrointestinal, cardiovascular, reproductive and nervous systems resulting in diverse metabolic and physiologic actions.

Tapsell et al., (2006) have discussed the health benefits of herbs and spices: the past, the present, the future. Pharmaceuticals are small-molecular-weight compounds consumed in a purified and concentrated form.

It is important to identify the bioactive substances in herbs and spices and also to know the properties.The antioxidant properties of herbs and spices are important to know the impact of oxidative modification of low-density lipoprotein cholesterol in the development of atherosclerosis. There is level III-3 evidence (National Health and Medical Research Council (NHMRC, 2013) wherein consuming a half to one clove of garlic (or equivalent) daily can have a cholesterol-lowering effect of up to 9%. There is level III-1 evidence that 7.2 g of garlic extract has been associated with anticlotting (*in-vivo* studies), as well as modest reductions in blood pressure (an approximate 5.5% decrease in systolic blood pressure). A range of bioactive compounds in herbs and spices have been studied for anticarcinogenic properties in animals. However, there is a challenge with the integration of knowledge to know any effects on humans within defined cuisines. Research on the effects of herbs and spices on mental health should distinguish between cognitive decline associated with ageing and the acute effects of psychological and cognitive function. There is level I and II evidence for the effect of some herbal supplements on psychological and cognitive function. There is very limited scientific evidence for the effects of herbs and spices on type 2 diabetes mellitus and the effect of ginseng on glycaemia.The use of food extracts is becoming a non-steroidal anti-inflammatory agent.There is an evidence that the use of ginger can reduce arthritic knee pain. Vegetable dishes are more appetising when prepared with herbs and spices. There are many metabolic diseases and age-related degenerative disorders which are closely associated with oxidative processes in the body in which the use of herbs and spices can be helpful. These developments are well underway through evidence-based frameworks for substantiating health claims related to foods. At present, recommendations are warranted to support the consumption of foods rich in bioactive components, such as herbs and spices. India is known as the land of spices and the 'spice route', has been a preferred trade destination for the Portuguese, British, Persians and other people from all over the world. In Indian cuisine no dish is prepared without spices. South Indian states are known for their use of tamarind and use of mustard seeds.The northern states are known for the use of cumin seeds. The eastern part is known for its preference to a mixture of 4 to 5 spices known as 'panch poran' is common.

Turmeric is well known as an antioxidant and cure for cough, cold and even cancer.

This is a very common spice of India. It is an antioxidant and has a property to Fight cancer, alzheimer disease and the inflammation of body joint. It also benefits the liver by cleansing it from toxins that come with excess alcohol.

Asafoetida

It is a strong and pungent spice.It usually changes the taste and smell of food.It is usually added to the daals while seasoning.It is helpful for Asthama, cough, bronchitis and digestion.It also helps during menstruation.

Cumin

Cumin known as zeera is used in lots of spicy foods. It is used in Indian, Middle Eastern, Eastern, Mexican, Spanish and Portuguese cuisines. Cumin is found in many curry powders, savory spice blends and latin/spanish spice blends. It smells spicy and the flavor is pungent, savory and slightly bitter. Cumin can be bought as seeds or ground into powder.It boosts the immune system, relieves nausea, stomach pain, cramps, digestion and diarrhea.It is used in Mexican/Spanish/Latin dishes such as chile, casseroles, enchiladas , spanish rice, mexican-spiced tofu scramble and jicama burgers. Cumin added to Indian dishes like tempeh in spicy onion curry and sophie dal gives a good taste. Moroccan dishes get added flavor from cumin.

Paprika

Paprika is a native of South America and provides umami flavor. It is also grown in Europe, particularly in Spain and Hungary. Spanish variety is mild while hungarian is very pungent. Paprika is a lovely red color and has a warm and sweet aroma. In hungarian dishes such as paprikash and goulash it is commonly used. It is used in sausages and in spice blends for seasoning flour used for making bread. To get a warm, smoky flavor, paprika is used after smoking in the recipes like coconut bacon, middle eastern vegan bake and in barbecue sauces. In India, paprika is used in tandoori dishes, giving the food a unique flavor and color.It is used on skin for measles, nerve pain, itchy skin caused by mites (scabies) and to treat pain. It is also used for insect bites. People inhale the oil to prevent falls, to help quit smoking and reduce cravings.

Red Chilli Flakes

Red chilli flakes, also known as crushed red pepper or red pepper flakes. These are made from a variety of red peppers like ancho, bell and cayenne. The fresh chillies are hot, the dried are more full-bodied and smoky. They are used in pickling, tomato sauces, soups, sausages, salads and pasta dishes. It is also used in the preparation of marinara sauce, spicy nachos and vegan pepper jack cashew cheese.

Chinese 5-Spice Powder

Chinese 5-Spice Powder is a blend which is a staple in Chinese cooking and has different varieties.The most common blend includes star anise, cloves, cinnamon, fennel and szechwan peppers. The aroma is dominated with the star anise. It is used in hoisin black bean burgers with spicy sesame sauce and in Chinese 5-spice potato latkes with plum hoisin dipping sauce. In ketchup it gives a good taste. Each of the ingredients in five-spice powder comes with its own set of health benefits. Cinnamon comes from the inner bark of tree that grows in Southeastern Asia. Cinnamon is known to have antioxidant and anti-diabetic properties and is also an antiseptic and anti-inflammatory agent. Its essential oils include eugenol, a fragrant compound that is an effective antiseptic. Other essential oils in cinnamon include cinnamaldehyde and linalool. The anticoagulant effects of cinnamaldehyde may prevent blood clots, which may help to reduce the risk of both strokes and heart attacks. These peppercorns come from a tree in the citrus family and are excellent Sources of essential oils, minerals and antioxidants. Their flavor comes from terpenes like limonene, linalool and cineol. They provide a boost to your digestion by increasing the secretion of gastro intestinal juices and enzymes, Star anise is the fruit of an evergreen plant that originated in southern China. Traditionally, it is used both as a spice and for its health benefits. It contains a plant-based compound called shikimic acid that is also found in ginkgo. When combined with quercetin, shikimic acid has been found to improve the functions of the immune system. Fennel seeds are produced by a plant in the parsley family that is grown in Europe as well as throughout Asia. It is a rich source of antioxidants like quercetin and of dietary fiber as well. It also contains health-improving essential oils like limonene and pinene, Cloves are the buds from an indonesian evergreen tree and are known to provide antioxidant, antiseptic and anti-inflammatory benefits. Among its essential oils are vanillin and acetyl eugenol but it also includes flavonoids like eugenin and kaempferol. It is known to provide high levels of vitamin A and can help to improve digestion by boosting the secretion of digestive enzymes.

Garam Masala

Garam Masala is a blend of spices from India. It contains the warming spices like pepper or cinnamon. Spices used in garam masala include coriander seeds, cumin, cardamom, mustard seeds, bay leaves, fennel, fenugreek, black peppercorns, cloves, mace and cinnamon.

Curry Powder

Curry powder is a spice mix from south asian cuisine. There are many varieties of curry powder and most include cumin, coriander, fenugreek, turmeric, chilli peppers and garlic. Some may have ginger, clove, cinnamon, fennel, cardamom, mustard seeds, nutmeg and black pepper.

Allspice is not a blend, is only one spice that smells like a combination of spices, particularly cinnamon, cloves, nutmeg and ginger. Allspice is indigenous to the rainforests of south and central America.It is also grown in Mexico and Jamaica. It is obtained from berries and it is used around the world for a multitude of purposes besides cooking. It could be whole or ground and smells aromatic as a combination of nutmeg, cloves, ginger and cinnamon. It is a warm spice that is sweet yet pungent and has a bit of peppery flavor. Allspice is found in Jerk dishes, marinades, pickling and mulling spices. It is used in baking and as a substitute for cinnamon, nutmeg or cloves. It is used in seasonal strawberry rhubarb pie, Jamaican rice and peas as well as jerk tofu.

Fennel

Fennel is both herb and a spice. It is a native of Mediterranean and an ancient plant known to the Greeks and Rome. It is grown in India, Australia, South America and the United States. It has a warm, sweet, aromatic aroma. It is a component of chinese 5-spice powder, curry powders and liquors. In Italy it is used to season sausages and spicy Italian vegan sausages. It is relished in sauteed spinach in fennel tomato sauce and also in marinara sauce. The fiber, potassium, folate, vitamin C, vitamin B-6, and phytonutrient content is coupled with its lack of cholesterol, all support heart health. Fennel contains significant amounts of fiber. Fiber decreases the risk of heart disease as it helps reduce the total amount of cholesterol in the blood. They are known to increase the milk supply in nursing mothers. They are used in many indian curries and pickles. Sugar coated "green supari mixtures containing 'saunf" are very common to eat after food.

Tamarind

It is from a bean like structure to provide a sour taste to many Indian curries. Dry Mango powder or Aamchur, is another very essential part of the Indian cooking which provides a tangy flavor to the dish. It is obtained after slicing unripe mangoes and dried in the sun for a long period of time and then ground into powder. It is also used in chaats around India. It is high in antioxidants, prevents atherosclerosis, rich in magnesium, promotes weight loss, good for nerve function, maintains blood pressure, promotes fluoride excretion, helpful for eyes, purifies the blood, is anti-diabetic and good for the skin.

Coriander

Coriander is an annual herb that is used in indian kitchen. Different parts of the coriander are edible but the fresh leaves and dried seeds are commonly used in cooking. Green coriander leaves after chopping are garnished on dishes after cooking while the dry coriander seeds are used either as whole or in ground form. Coriander roots are used in various soups due to the presence of an intense flavour. It has antioxidant and anti inflammatory activities. The leaves or powder are used against digestion problems including stomach upset, appetite loss, hernia, nausea, diarrhea, bowel spasms, and intestinal gas. It is also used to treat measles, hemorrhoids, toothaches, worms and joint pain as well as infections caused by bacteria and fungus.

Ginger

Several studies have investigated ginger's effects on the gases that form in the intestinal tract during digestion. Some research indicates that enzymes in ginger can help the body break up and remove the gases, providing relief from any discomfort. Ginger also appears to have beneficial effects on the enzymes trypsin and pancreatic lipase, which are important for digestion. In addition, ginger may help the movement of food in the digestive tract preventing constipation. Some research indicates that ginger can help alleviate morning sickness and relieve nausea following cancer treatment. The effects of ginger root powder were studied on nausea who underwent chemotherapy. The results have shown that it reduces nausea. Also a daily dosage of 1, 500 milligrams (mg) of ginger extract reduces nausea, gastrointestinal problems and to recover from a cold or flu. A study was done on the effects of fresh and dried ginger on one respiratory virus in human cells.The results indicate that fresh ginger may help to protect the respiratory system, while dried ginger did not have the same impact. A daily dosage of 2 grams (g) of raw or heated ginger reduces exercise-induced muscle pain by about 25%.A review has shown that ginger may help in reducing dysmenorrhea, pain either before or during menstruation, in treating inflammation caused by osteoarthritis, cardiovascular disease, antiplatelet activity and in diabetes. It can reduce various types of oxidative stress. It has also been found that people who had consumed ginger did not have negative changes in healthy colon tissue. Ginger (*Zingiber officinale* Roscoe *Zingiberaceae*) helps in the treatment of peptic ulceration due to its action as a thromboxane synthetic inhibitor (Haniadka et al., 2013). Maximum activity has been found in the fraction containing a-zingiberene, p sesquiphellandrene bisabolene and curcumene (97.7% inhibition at 125 ppm; Kochhar, 2008).

Cardamom (*Elettaria cardamomum*)

It is a spice in small pod with black seeds inside. The seeds are sweet and savory in taste. It is used in desserts, hot and spicy dishes, as well as aromatic beverages, coffees and teas. [http://www.imedpub.com/articles/herbal-medicines-possible-risks-andbenefits.pdf It originated in India, Nepal and Bhutan.It is traditionally considered as a therapeutic herb and has application in Ayurveda (the ancient Indian science of medicine and lifestyle) and in traditional Chinese medicine[https://www.ncbi.nlm.nih.gov/pmc/articles/PMC3170500/pdf/nihms307509.pdf]. It is used as a remedy or constipation, colic, diarrhea, dyspepsia, vomiting, headache, hypertension (high blood pressure), epilepsyandcardiovascular diseases[https://ndb.nal.usda.gov/ndb/foods/show/02006]. The anticancer potential, [https://www.ncbi.nlm.nih.gov/ pmc/articles/ PMC4827078/pdf /EXCLI-12-1043.pdf] a combination of cardamom and cinnamon was found to be helpful in reducing the risk of colorectal cancer by 48%, by enhancing antioxidant activity in the body [http://www.jmb.or.kr/journal/viewJournal.html?doi=10.4014/jmb.1605.05056].It is protective against heart inflammation [https://www.ncbi.nlm.nih.gov/pmc/articles/PMC4661898/pdf/ijms-16-26040.pdf]. The benefit for blood pressure and hypertension has been proved [http://nopr.niscair.res.in/bitstream/123456789/19002/1/IJEB%2037%283%29%20 238-242.pdf [http://www.eurekaselect.com/146591/article. In an animal experiment the antioxidant enzymes from cardamom were found to protect the heart from oxidation and to control cholesterol levels despite diets high fat or cholesterol https://onlinelibrary. wiley.com/doi/abs/10.1002/jsfa.8165, https://www.ncbi.nlm.nih.gov/pmc/articles/ PMC5557534/ pdf/ 12944 _ 2017_Article_539.pdfand reduces obesity [https://www.ncbi.nlm.nih.gov/pubmed/28603121]. Laboratory studies have [https://onlinelibrary.wiley.com/doi/abs/10.1002/jsfa.8414] found cardamom to contribute towards anti-diabetic effects and for the regulation of glucose and insulin metabolism.It has anti-inflammation and oxidative stress in hyperlipidemic, overweight, and obese pre-diabetic women [https://books.google.co.in/books?id=nZXMC7-_BpMC&pg=PA57&lpg=PA57&dq= and anti-epressant properties [The essential oil is used in anti-stress aromatherapy [https://www.researchgate.net/publication/312151655_The_effect_of_Elettaria_cardamomum_extract_on_anxiety-like_behavior_in_a_rat_model_of_post-traumatic_stress_disorder]. The extract can reduce the anxiety symptoms in test animals [https://www.ncbi.nlm.nih.gov/pmc/articles/PMC5775110/pdf/main.pdf]. In a study on mice with cardamom during pregnancy, their offsprings showed enhanced learning and memory behavior [https://www.sciencedirect.com/science/article/abs/pii/S1043661896900675?via%3Dihub. It also helps in gastrointestinal discomforts like acidity, flatulence and stomach cramps.

They have potential against *Helicobacter* pylori, the bacterium responsible for stomach ulcers. Cardamom oils inhibit the growth of dangerous microbes that cause food poisoning https://www.ncbi.nlm.nih.gov/pmc/articles/PMC6278479/pdf/molecules-23-02818.pdf, https://www.ncbi.nlm.nih.gov/pmc/articles/PMC5048087/pdf/fmicb-07-01580.pdf also active against antibiotic-resistant infection (https://www.mdpi.com/2079-6382/8/1/8; https://www.sciencedirect.com/science/article/pii/S0753332216329535).

Certain compounds have antioxidant and anti-inflammatory properties [https://www.ncbi.nlm.nih.gov/pmc/articles/PMC3353705/.

Cardamom has been used in ayurveda and traditional chinese medicine for dealing with dental problems [https://core.ac.uk/download/pdf/60543480.pdf)., against asthma and other respiratory issues and protection against inhalation injury in the lungs of mice [https://www.researchgate.net/publication/7655083_Cardamom_extract_as_inhibitor_of_human_platelet_aggregation].In traditional therapies like aromatherapy, cardamom has been said to improve blood circulation to the lungs, the aromatherapy can reduce stress and oxygen circulation in the body [https://www.researchgate.net/publication/312151655_The_effect_of_Elettaria_cardamomum_extract_on_anxiety-like_behavior_in_a_rat_model_of_post-traumatic_stress_disorder].It can protect the blood from clogging vessels .Traditionally used as a remedy for nausea[http://nmj.umsha.ac.ir/browse.php?a_id=1280&sid=1&slc_lang=en]. Morning gargling in warm water could prevent sore throats [https://www.researchgate.net/publication/321573429_Processing_and_Medicinal_Uses_of_Cardamom_and_Ginger-A_Review]. It has been used against impotency as it enhances the blood circulation.Cardamom have been used as a remedy for urinary tract diseases [http://books.google.co.in/books?id=WJP7tgjzE_oC&pg=PA64&lpg=PA64&dq=cardamom+hiccups&Source=bl&ots=NsgzAII4lj&sig=Q1VoRwLIwBaG1ZDqvj2oYWXIJ0w&hl=en&sa=X&ei=1LBFUN1qieSsB-HwgdgF&ved=0CC8Q6AEwAA#v=onepage&q=cardamom%20hiccups&f=false] and kidney disorders such as cystitis, nephritis and bleeding, as well as gonorrhea.Hiccups are spasms in the tissue just below the lungs, cardamom infusion made by boiling the spice powder in water for an extended time to concentrate the active components has been found to be beneficial [https://www.sciencepubco.com/index.php/ijbas/article/view/5435]. The seeds are chewed as a breath freshener, and the essential oil is used in chewing gum [https://core.ac.uk/download/pdf/60543480.pdf].High levels of cholesterol, triglycerides and liver enzymes can be controlled with cardamom extracts [https://www.ncbi.nlm.nih.gov/pubmed/27246748. The extract can reduce the risk of fatty liver disease and prevents liver enlargement [https://www.ncbi.nlm.nih.gov/pmc/articles/PMC3975278/pdf/1749-8546-9-10.pdf].

It belongs to the family of Zingiberaceae [http://www.indianspices.com/spice-catalog/cardamom-small-1]. The two main varieties are Elettaria, green or true cardamom and Amomum, the black, white or red cardamom.It is an origin of India, now it is available in tropical regions of Asia, including India, Sri Lanka, China, Bhutan, Vietnam, like Malaysia, Korea, Japan, Guatemala (Central America) and Tanzania (Africa).The Spices Board of India [http://www. rediff.com/money/2007/may/30farm.htm]. has recognized three varieties of cardamom, the Malabar, Mysore and Vazhukka varieties. Another variety is Njallani [https://food52.com/blog/16041-for-middle-eastern-style-coffee-add-cardamom], which is popular along with the three normal varieties. http://naturalingredient.org/wp/wp-content/uploads/cardamom-benefits-_-information.pdf.

Cloves

It has originated in Indonesia and are dried flower buds from the clove tree. It is widely used in soups, stews, meats, sauces and rice dishes, traditional medicine and health benefits.The clove oil is a natural way for maintaining oral health due to its effect on plaque, inflammation, gingivitis and bacteria in the mouth. Clove extract and a component called nigericin, reduced insulin resistance in mouse muscle cells. It has shown improvements in glucose tolerance. According to Today's Dietitian, 1/2 teaspoon of ground clove is said to contain more antioxidants than 1/2 cup of blueberries. In a study scientists found that clove extract was able to slow the growth of multiple types of human cancer cells. Clove extract also increased cell death in colon cancer cells.Tumors grew significantly less in the mice treated with clove extract compared to those in the control group. It has been found that clove essential oil and ethanol extract of clove are toxic to breast cancer cells. In a study researchers found that clove extract reduced the incidence of obesity resulting from a high-fat diet. According to the National Center for Biotechnology Information, multiple hazards are associated with clove oil. It may cause skin, eye, respiratory irritation, an allergic reaction on the skin. It is also flammable and could be fatal if it is swallowed and goes into the airway. In an incident with a 2-year-old child who drank 5 to 10 ml of clove oil experienced multiple medical problems, including coma, liver damage and problems with blood clotting. It may increase the chance of bleeding or increase the body's response to warfarin. Ground cloves are used as flavoring. In the recipe for gingerbread cake with cream cheese frosting, ground cloves are used. It is used to make masala chai, which is a mixture of tea, spices, and milk popular in India and Pakistan. Cloves are used in savory dishes as in this recipe for baked chicken with artichokes, cinnamon and preserved lemons and this one for slow-cooker braised beef with carrots and turnips.According to the

United States Department of Agriculture National Nutrient Database, 2.1 grams (g) or 1 teaspoon of ground cloves contains:6 kilocalories (kcal), 0.13 grams (g) of protein, 0.27 g total fat, 1.38 g carbohydrate and 0.7 g fiber.The same amount of ground cloves also provides 1.263 milligrams (mg) of manganese, which is 63 percent of the Daily Value (DV) for manganese(Megan 2018).

Saffron

It is the most expensive spice in the world pricing of 1 pound (450 grams) between 500 and 5, 000 U.S. dollars.It is due to the labor-intensive harvesting method making the production costly. As it is harvested by hand from the *Crocus sativus* flower, commonly known as the "saffron crocus." The term "saffron" applies to the flower's thread-like structures or stigma. It was originated in Greece, where it was revered for its medicinal properties. Traditionally it is eaten to enhance libido, mood enhancer and improve memory.Saffron has antioxidant activity against free radicals and oxidative stress which include crocin, crocetin, safranal and kaempferol. The carotenoid pigments give the red color. They also have antidepressant properties, protect the brain cells against progressive damage, improve inflammation, reduce appetite and aid weight loss. Kaempferol found in the flower petals provides health benefits like reduced inflammation, anticancer properties and antidepressant activity. It can treat symptoms of mild-to-moderate depression.A study has shown that taking 30 mg of saffron (petals /stigma) daily was as effective as Fluoxetine, Imipramine and Citalopram which have been used conventionally.It has shown to kill colon cancer cells or suppress their growth, while leaving normal cells unharmed to skin, bone marrow, prostate, lung, breast, cervix, and several other cancer cells.Premenstrual syndrome (PMS) is a term that describes physical, emotional, and psychological symptoms occurring before the start of a menstrual period and saffron has helped in it.

Even smelling of saffron for 20 minutes has shown to reduce PMS symptoms like anxiety and has owered the levels of the stress hormone cortisol. Studies have shown that saffron has aphrodisiac properties in people taking antidepressants.Studies have shown that taking saffron significantly improved erectile function, libido and overall satisfaction. It may help in curbing the appetite. It may lower blood cholesterol and prevent blood vessels and arteries from clogging , may lower blood sugar levels and raise insulin sensitivity, can improve eyesight in adults with AMD and protect against free radical damage, which is linked to AMD and can help in Alzheimer's disease.In small doses, it is added to savory dishes, like paella, risottos and rice dishes.The threads are soaked in hot but not in boiling water. It is then added to the recipe for a richer flavor. Usually it is safe and has no side effects. Up to 1.5 grams of saffron per day can be taken as a dietary supplement. High doses of 5 grams or more can have toxic effects. It is advisable that pregnant women avoid high doses as it can cause miscarriage(Ryan Raman, 2019).

Cinnamon

The inner bark of tree Cinnamomum gives this spice.It has been used since ancient times for its therapeutic uses. Two types of cinnamon are Ceylon cinnamon known as "true" cinnamon and Cassia cinnamon is a common variety generally referred as "cinnamon."It is obtained by cutting the stems of cinnamon trees where the inner bark is extracted and the woody parts are removed. After drying it forms strips that curl into rolls, called cinnamon sticks. These sticks can be ground to form cinnamon powder.It has a distinct smell and flavor due to the oily part, which is very high in the compound cinnamaldehyde which is responsible for good health and metabolism.It has antioxidants, such as polyphenols.It can be used as a natural food preservative also.It helps the body to fight against infections, repairs the tissue damage, reduces the risk of heart disease, has beneficial effects on blood markers, reduces the levels of total cholesterol and increases good HDL cholesterol levels.It helps to reduce insulin resistance by lowering the blood sugar levels. It has been shown to decrease the amount of glucose that enters the bloodstream after a meal.The effective dose is in between 1–6 grams per day. Two compounds in cinnamon inhibit the buildup of a protein called tau in the brain, which causes Alzheimer's disease. In a study with mice having Parkinson's disease, cinnamon was found to help in protecting neurons, normalized neurotransmitter levels and improved motor function. Animal studies, suggest that cinnamon extracts may protect against cancer.It helps in the formation of blood vessels in tumors and is toxic to the cancer cells causing cell death.Cinnamaldehyde, the main active component of cinnamon, helps to Fight various kinds of infection, treats respiratory tract infections caused by fungi and can inhibit the growth of certain bacteria, including *Listeria* and *Salmonella*.The antimicrobial effects of cinnamon also prevents tooth decay and reduces the bad breath.It helps against HIV-1, the most common strain of the HIV virus in humans. Significant amounts of a compound called coumarin, are present in the Cassia variety have health benefits, but may cause problems in large doses.

Fenugreek Seeds or Methi Seeds

These are small, flat, squarish, brownish-beige color seeds used in curries. Due to a slightly bitter flavor they are not used in large quantities. They are especially used in fish curries. They are added to curry powders, green leaves are used in Indian cooking giving a bitter taste which is quite piquant but acceptable. Balances the cholesterol levels, soothens the upset stomach and digestive problems, reduces menstrual cramps, reduces appetite, reduces fat mass, maintains liver and kidney health, soothes muscle pain and reduces fever.

Sesame Seeds or Til

These are used in many masalas and some chutney for the particular flavo and healthful properties. It helps in Diabetes, blood pressure, arthritis, healthy skin and bones. It contains high amount of folic acid and is a good source of vitamin B complex and gives a shine to the hair.

Nutmeg or Jaiphal is usually used in powdered form. It is used in flavoring Indian sweets. It is used in savory dishes and in making Garam Masala. It is helpful for insomnia, irritability and nervousness. The tree is native to Asia, the leaves are small and very shiny and are used fresh. The leaves are fried in oil, until crisp, before the preparation of a curry.

Sambar masala contains dry red chillies : 1/2 kg, dhania/ coriander seeds : 1/2 kg , urad dal : 1/4 cup , channa dal : 1/4 cup, toor dal : 1/4 cup, rice grains : 1/4 cup, black pepper corns : 50 grams, cumin seeds : 50 grams, mustard seeds : 30 grams, fenugreek seeds : 10-20 grams, katti perungayam / asafoetida : 20 gram, fresh curry leaves : a big bunch. in 1 teaspoon of oil add the sun dried ingredients one by one separately and dry roast them in heavy bottom pan.

Rasam Powder contains dry roast all the ingredients in a medium low flame till they get evenly browned, cool and grind into a fine powder.

Pav Bhaji Masala :Black cardamoms, 4; coriander seeds 1/4 cup, cumin seeds 1/8 cup, whole black pepper 1/8 cup, fennel seeds 1 and 1/2 tbsp, whole dry red chillies 10, cinnamon 10 inch pieces, dry mango powder, 2 tbsp. For the preparation, dry roast all the ingredients till golden brown (except dry mango powder) in low medium heat, roast it till nice aroma is released. Then add mango powder and roast for a minute. Cool down completely and then grind it to a fine powder.

Devarajan and Mohanmarugaraja, (2017) have written a very good review on the functional properties of rasam. Dietary spices may influence gastric emptying, gastrointestinal motility, secretion of gastric acid as well as intestinal bicarbonate, bilopancreatic secretions, absorptive processes and bacterial microflora (Little et al., 2010). The effect of spices on gut mucosa, gastrointestinal reflexes and enteric nerves as well as systemic autonomic response have a large influence on carbohydrate, protein and fat absorption in different ways (Browning and Travagli, 2014). Changes in mucus barrier, gastroprotective effects, enterocyte, change in villous height and cytoprotective adaptations are known to occur with intake of spices or spice principles (Kochhar, 2008). Spice intake is associated with gut influencing secretion, motility, absorption and change in microflora (Kochhar, 2008). Capsaicin is the pungent active principle of red chilli or paprika and it has been shown to cause gastric mucosal oedema and hyperemia and decrease the gastric acid output (Bevan and Szolcsányi, 1990). An analogue of capsaicin as well as dry capsaicin powder help in gustatory sweating and in salivation (Basith et al., 2016). It protects the gastric mucosa against damage induced by pyloric ligation, ethanol and acidified asprine (Holzer et al., 1969).Intragastric capsaicin promotes the release of vasodilator substances like nitric oxide (NO) (McCarty et al., 2015), it is also associated with release of calcitonin gene related peptide (CGRP) which via formation of NO may cause both a prevention of mucosal injury and modulation of gastrointestinal motility (O'Neill et al, 2012). Chilli causes dyspepsia in patients and are advised to avoid its use in sensitive patients (Marotta and Floch, 1991). Nevertheless, epidemiological and clinical data suggests that chilli ingestion may have a beneficial effect on human peptic ulcer disease (Kang et al., 1995). For instance chilli has a gastroprotective effect on aspirin-induced gastric mucosal injury in humans (Satyanarayana, 2006). Chilli ingestion does not cause macroscopic damage in humans. Long-term chilli intake (360 mg daily for 4 weeks) protected against hemorrhagic shock-induced gastric mucosal injury (Brzozowski et al., 1996). An analogue of capsaicin, known as resiniferatoxin has antiulcer activity. It acts on vanilloid (capsaicin) receptors. Paprika, pepper and cinnamon increased gastric acid secretion in men and promoted histamine induced ulcer formation in dogs, whereas mustard decreased acid output (Szabó et al., 2013). The analysis done by video-endoscopy did not demonstrate any visible gastro duodenal damage on the intake of spicy foods (Kochhar, 2008). To study the effect of spices on gastric acid secretion the stomach of pentobarbitone-anesthetized rats was perfused at 0.15 ml/min with aqueous extracts (10% w/v) of red pepper (*Capsicum annuum*), fennel (*Foeniculum vulgare*), omum/ajwan (*Carum copticum*), cardamom (*Elettaria cardamomum*), black pepper (*Piper nigrum*), cumin (*Cuminum cyminum*) and coriander (*Coriandrum sativum*) or acetylcholine. All the spices tested increased acid secretion. Atropine abolished the acid secretion induced by acetylcholine and significantly reduced acid induction by red pepper, omum and coriander, but not that by fennel (Govindnarayana, 1991). The spices tested increased the gastric acid secretion (Raybould et al., 1988). Licorice, the root and rhizome of different varieties of Glycyrrhiza glabra, has been used as medicine for antiulcer activity (Ahmad et al., 2016). Licorice has an anti ulcer activity by stimulating the gastric mucus production, enhancing the rate of incorporation of various sugars into gastric mucosal glycoproteins and by promoting the mucosal cell proliferation. It has been found that it inhibits the mucosal cell exfoliation, inhibits prostaglandin degradation, increases the release of PGE2 and reduces the formation of thromboxane B2 7. It also helps in the regulation of DNA and protein synthesis in gastric mucosa.

The spices like columbo weed, long pepper, parsley, tarragon, nutmeg, yellow berried nightshade, threadstem carpetweed, sage and cinnamon all had bactericidal activities against *Helicobacter pylori* and those of turmeric, borage and parsley were able to inhibit the adhesion of *H. pylori* strains to the stomach section (O'Mahony et al., 2006). In another study the *in vitro* susceptibility of extracts of spice and food plants from Thai traditional medicine inhibited the growth of *H. pylori* (Mahady et al., 2005). Therefore the incidences of gastric cancer are few in Asian countries (Mahady et al., 2005). Spices like pepper, asafetida, aloe, ocimum and eugenol are known to have hypoglycaemic action (Kochhar, 2008). Fenugreek seeds (*Trigonella foenumgraecum*), garlic (*Allium sativum*), onion (*Allium cepa*), and turmeric (*Curcuma longa*) have been experimentally documented to possess potential to function as antidiabetic agents (Viles et al., 1995). Cumin seeds (*Cuminum cyminum*), ginger (*Zingiber officinale*), mustard (*Brassica nigra*), curry leaves (*Murraya koenigii*) and coriander (*Coriandrum sativum*) have also been reported to have hypoglycaemic effects (Madkor et al., 2011). Both *Murraya koenigii and Brassica juncea* showed significant hypoglycemic action in experimental rats (Khan et al., 1995).The stimulatory effect of capsaicin on serum insulin may be via capsaicin induced beta adrenergic activity on B cell receptors in pancreatic islets (Emir TLR, editor.Boca Raton, 2017). Many enzymes of the liver get affected by spices both *in vitro* cell culture systems as well as *in vivo* in experimental animals (Kochhar , 1996). An experimental study on some physiological effects of dietary spices has been conducted by a thesis submitted to AIIMS for the Degree of Doctor of Philosophy(1996).

Other Spices commonly used around the Globe

Food is nothing without spices, and every country has its own preferred ones. It has been observed that cumin is the most popular spice in the world and coriander (cilantro) is the most commonly used herb. However, in Europe and Africa, garlic is commonly used among all dishes. Oregano is found to be common in the Mediterranean regions. In a review (Henrietta et al., 2017) have discussed the role of spices in nutrition and health. Apart from the use as organoleptic enhancers in food the spices are finding beneficial utilization in healthcare. Spices like *Piper guineense*, *Afromomum melegueta* and *Tetrapleura tetrapetra* are commonly used in the Southern part of Nigeria. The nutritional quality, phytochemistry and medicinal properties of these spices have been studied. They have different varieties of flavor, color and aroma and provide nutrients to foods (Mann, 2011, Kaefer and Milner, 2011). Many spices have pharmacological and medicinal properties (Newman and

Cragg, 2012). Some of their constituents can be used as defense chemicals to repel insects, snails and other animals and helps in killing parasitic fungi (Adeyemi, 2011). Presence of allelochemicals in some plants helps in inhibiting the growth of other plant species around them (Linhart et al., 2015). They constitute almost 40% of prescribed drugs till date (Mann, 2011).Spices and herbs are revered for their potential health attributes. They can control chronic diseases like cancer, diabetes and cardiovascular diseases (Kaefer and Milner, 2011). Many studies have proved this fact (Kochhar, 2008; Krishnaswamy, 2008; Iyer et al., 2009).Use of phytomedicine for treatment and management of diseases has been worked out well (Sigh, 2007; Dolui and Segupta, 2012). New areas in the fields of nutraceuticals and functional foods have been opened up. Curcumin (turmeric), eugenol (clove), and capsaicin (red pepper), have proved to control oxidative stress in cells due to their antioxidant properties and their capacity to block the production of oxygen radicals in aerobic metabolism and interfering with signal transduction pathways (Rubió et al., 2013; Srinivasan, 2014). Polyphenols from ginger (*Zingiber officinale*) and turmeric (*Curcuma longa* L.) have shown free radical scavenging properties (Scalbert and Williamson, 2000). *Piper guineense* Schum and Thonn (Piperaceae) are commonly known as African black pepper, 'uziza' in Igbo South East, and 'iyeree' in Yoruba south Western Nigeria. The plant is also known as Ashanti pepper, Benin pepper, Guinea pepper and false cubeb in other parts of Africa. More than 700 species of this plant are found in the tropical and sub-tropical Africa (Besong et al., 2016). The spheroid seeds, are semi-cultivated in Nigeria (Klin-Kabari, 2011). The plant is used for culinary, medicinal, cosmetics and insecticidal purposes (Martins, 2013; Anyawu and Nwosu, 2014).In the South Eastern parts of Nigeria, the seeds are prescribed for women after childbirth to enhance uterine contraction enabling expulsion of the placenta and other remains from the womb (Ekanem et al., 2010). It helps in treating rheumatic pains and also as an anti-asthmatic agent (Sonibare and Gbile, 2008) and in decreasing obesity (Mba, 1994). The oil distillate from the seeds is used in perfumery and for making soap. It is possible to regulate menstrual cycle and female infertility by the leaves (Iwu, 2014). The root and fruits are helpful in sexually transmitted diseases like gonorrhea and syphilis (Iwu, 2014). Alkaloids act as CNS stimulant and powerful analgesics (Ashok and Upadhyaya, 2012).Saponins have been reported to have antimalarial effect (Besong et al., 2016). Flavonoids have been reported to possess antioxidant, anti-inflammatory, antitumor, antiallergic and antiplatelet activity (Pal and Verma, 2013). The essential oils consists of compounds like dillapiol, piperine (5%–8%), elemicine, myristicine (10%) and safrole that show bactericidal and antimicrobial activity on some microorganisms (Klin-Kabari et al., 2011). The amount of β-caryophyllene has shown anti-inflammatory properties

(Issac, 2012). *P. guineense* has crude protein, carbohydrates, fat, vitamins and minerals (Nwakwo et al., 2014). The calcium content of *P. guineense* is high (Omotayo et al., 2013). *P.guineense* contains vitamin A, C and E and traces of vitamin B_1, B_2 and B_3. The presence of vitamin E points to the antioxidant capacity of the plant (Ogbunugafor et al., 2011). Good amount of vitamin C indicates that it supports the formation of healthy gum, teeth and the healing process.The effect of the leaf extract on uterine contraction has been studied by Udoh et al. (1999). Cholinergic activity of the leaf extract has been found on uterine muscle (Udoh et al. 2012).

Mbongue et al. (2005) investigated the effect of dry fruits of *P. guineense* on the reproductive functions of adult male rats and found that the administration of the aqueous extract of fruits of the plant had a positive impact on the male reproductive functions by stimulating the secretions of the testes, epididymis and seminal vesicles.Uhegbu et al. (2015) found that the fruit extract significantly decreased the liver enzymes ALT and ALP in the rat. They suggested that the seed extract might possess hepatoprotective ability. Anyawu and Nwosu (2014) found that the extracts of the leaves of the plant had antimicrobial activity against the bacteria *S. aureus*, *E. coli*, *P. aeruginosa*, *B. subtilis* and the fungi *C. albicans* and *S. cerevisiae.* Kabiru et al. (2016) have reported a reduction in parasitemia in a dose-dependent pattern.The anticancer activity of leaf extract of *P. guineense* against human myeloid leukemia (HL-60) cell line has been reported by Iweala et al. (2015).Olawuyi et al. (2013) showed the anti pesticide activity. Phytoremediation effect of *Piper guineense* on the levels of polycyclic aromatic hydrocarbons (PAHs) was evaluated and Etim et al. (2014) found that the plant could be good for phytoremediation of phenanthrene, chrysene benzo(b)fluoranthene, dibenzo(a, b)athracene, and indeno pyrene.The seed of *A. melegueta* is used in different African cultures as a spice and medicine as a local remedy for stomach ache, snakebite, diarrhea, cardiovascular diseases, diabetes and inflammation (Ilic et al., 2010). The seeds of *A. melegueta* have high amounts of carbohydrates, crude fibre and minerals (Dike and Ahamefula, 2012; Echo et al., 2012; Alaje et al., 2014). Kuete et al. (2011) have shown that the extract of *A. melegueta* to have significant inhibitory activities (IC_{50} value above 10μg ml^{-1}) on human pancreatic cancer (MiaPaca-2) and leukemia (CCRF-CEM) cell lines. It can also treat diabetes (Ogbera and Ekpebegh, 2014). In a study by Adefegha et al. (2017), the oil from *A. melegueta* was shown to inhibit α-amylase and α-glucosidase activities. Mojekwu et al. (2011) reported the reduction in blood glucose levels.The seeds of *A. melegueta* has been reported to possess anti-inflammatory properties (Ilic et al., 2014). In a study by Kamtchouing et al. (2002) in male Wistar rats administration of 115 mg kg^{-1} of *A. melegueta* daily for 8 days led to an increase in the penile erection index (PEI).The seed extract of *A. melegueta* prevented diarrhoea, intestinal fluid secretion and gastrointestinal transit time induced by castor oil in rodents. *Tetrapleura tetraptera* Schumach and Thonn Taub (Mimosaceae) a deciduous plant grows on the fringe of the West and Central African rainforest zone (Uyom et al., 2013; Adesina et al., 2016). The fruit has a pungent aromatic odor, with insect-repellent property (Adetunji, 2007). According to Orwa et al. (2009) every part of the plant is useful in industries. The fruit pulp is used in flavouring food, the hard heartwood and white sapwood are used in timber, the tannin is used as a dye and also in medicine, extracts of the leaves, bark, roots and kernel. The fruits and flowers are used as perfumes (Ngassoum et al., 2001).In a study done by Ekwenye and Okorie (2010) has shown that alkaloids and saponins were present in the pod.Udourioh and Etokudoh (2014) analyzed the essential oil and fatty acid composition of the dry fruits.The presence of the short chain fatty acids like omega-6 and omega-3 makes it beneficial for human health. In a study done by Abil and Elegalam (2007) it was found that the fruits of *T. tetraptera* contained Ca, P, K, Mg, Zn, and Fe with Zn.The isolated compounds showed sedative, hypotensive, molluscidal, CNS depressant, anti-inflammatory, antimicrobial, wound-healing, contraceptive, analgesic, hypoglycemic, antioxidant, hypolipidemic, antimalarial, muscle-relaxant, anticonvulsant, hypothermic and anxiolytic effects (Adesina et al. 2016). Ojewole and Adewunmi (2004) reported that the aqueous extract of the fruit showed anti-inflammatory activity in egg albumin-induced pallet edema in rats. The anti-inflammatory activity of *T. tetraptera* is linked to the hentriacontane compound it contains. Adesina et al. (2016) found that the extract decreased the blood glucose level in the animals. It also has analgesic effects, the aqueous fruit extract delayed the onset of seizures (Ojewole, 2005). Achi (2006), Ekwenye and Okorie (2010), Aboaba (2011) and Oguoma et al. (2015), have studied the antibacterial activity of the fruit of *T. tetraptera* and found it effective against human pathogens like *Salmonella typhi, Bacillus subtilis, P. aeruginosa*, *Escherichia coli, Shigella* spp. and *Staphylococcus aureus*. The extract has activity against these fungi pathogens namely, *A. niger* and *P. notatum* (Igwe and Akabuike, 2016). Takahashi et al. (1995) has given a report on the antiviral activity of the plant.The fruit extract was also evaluated for its antiplasmodial activity by Okokon et al. (2007). Adesina et al., (2016) have looked into molluscicidal activity and found that all parts of the plant possess it.

7

Traditional Foods Around the World and the Role in Nutritional Security

With increasing food production, availability and accessibility the global agro-industrial food system has improved.However, malnutrition, biodiversity loss and climate change are newer challenges. The best solution to this is the utilization of indigenous and traditional crops which will bring a transformative change in the food system. There is a need to look into the development of smallholder producers. The agro-industrial food system has improved the food supply and has made way for significant trade-offs with agro-biodiversity, dietary diversity, environmental sustainability and socio-economic stability mainly amongst the rural poor. Now is the time to mainstream the underutilized indigenous and traditional crops into the food system. This will provide opportunities to develop a sustainable and healthy food system.It will also help towards the societal goals which include employment creation, wellbeing and environmental sustainability.

It is the role of the researchers to take it to the policy-makers. Also, the policy-makers must acknowledge the divergence in the existing policies and should have a policy of a food system which includes smallholder farmers. Food security, as defined by the FAO, is a situation that exists when all people, at all times, have physical, social and economic access to sufficient, safe and nutritious food in order to meet their dietary needs and food preferences for an active and healthy life. It has been found that it is the traditional foods from local environments (plants and animals) which form a part of the diet. This is linked to specific regional ecosystems and offers sustainable food and nutrition security(Johns et al, 2013; Johns and Eyzaguirre, 2006). Frewer et al., (1997) gave methodological approaches to assess risk with food related hazards.

Recently, dietary shifts are happening globally, from consumption of traditional diets, which usually has more nutrients.It also provides more energy.This dietary shift, due to globalisation and urbanisation, is known as 'nutrition transition'.The 2013 agricultural survey showed that the number of farmers dropped from 121, 766 in 2012 to 119, 134 in 2013 (Statistics Botswana, 2015). The sector does not produce enough food to feed the nation and consequently, the population relies on food imports and purchase of foods (Statistics Botswana, 2015;Raboloko, 2016;Acquah et al, 2014).The national food security policy usually focuses on food imports in most of the countries. Supermarkets are the major food distributors of imported, processed and packaged foods(Food and Agriculture Organization of the United Nations (2010) http://www.fao.org/3/a-i1951e.pdf. Many households depend on the super markets.However, the heavy reliance on imported and processed foods is having an adverse effect on the diverse local, traditional and indigenous diets and food security. This is also giving rise to many diseases.The country's prevalence of undernourished people (with energy intake below the minimum dietary energy requirement) stated by the FAO in 2015 was 28·7% , however it has been found to be very low in most of the countries. Not many households are utilizing the traditional indigenous foods (TIF) as people are not aware of their importance in achieving food and nutrition security.With increased 'supermarketisation' globally. However, it is important that the local foods are supplied to the supermarkets and may also be imported from the surrounding regions and countries (Anand et al., 2015). The modern foods are energy-dense and are poor in nutrients. This leads to the risk of obesity and chronic diseases (Sharma et al., 2009; Erber et al., 2010). The dietary shift towards modern foods may pose future public health challenges. Workers have found that TIF are associated with food security in terms of access and dietary diversity. In 2012, the FAO (Food and Agriculture Organization of the United Nations, 2012) observed that traditional foods are crucial for food and nutrition security, especially for poor families in rural, peri-urban and urban areas, because of their nutritional benefits and affordability. Galvin et al., (2015) gave the nutritional status of Maasai under change.

It is the traditional and indigenous foods that can contribute to food security the eradication of hunger and poverty, as well as the prevention of diseases. Indigenous fruits, vegetables, tubers and roots can play a major role in enhancing the quality of diets and improving food and nutrition security.There is sufficient evidence from studies from South Africa and the rest of the world that indigenous foods are consumed by many households, especially in areas where they are available, and thus can play an important role in alleviating food insecurity. Many rural communities have access to traditional crops that are rich in micronutrients and can serve to eliminate food insecurity. As these are diverse in nature they have the ability to augment the nutrient composition of family diets.This will help in household food security and for the alleviation of hidden hunger which comes due to the lack of dietary diversity and is linked to poor consumption of fruit and vegetables in general.Underutilised indigenous crops provide an opportunity for incorporation of alternative food sources into rural food systems and can be good. Sources of nutrients augmenting better health.It has been found that indigenous foods and dietary diversity in an ecosystem may be very helpful. It is the food insecurity which is a major issue in developing countries and leads to hunger, malnutrition, negative effects on health and poor quality of life. It also affects the social and economic development of human beings.

Recently, in a survey done by South African National Health and Nutrition Examination reported that only 45.6% of the South African population had the food and nutrition security. The largest percentage of participants who experienced food insecurity was found in urban informal (32.4%) and rural formal (37.0%) localities. Bruce (1958) looked into the economies of staple food of western tropical Africa. Addressing food insecurity needs policies to know indigenous foods with their potential nutritive value, especially those harvested in impoverished communities.The use of indigenous foods have declined due to the nonavailability in modern and industrialised markets and a lack of investment in research and development. Moreover, there is not enough information available regarding their nutritional value.If at the global and national level, efforts are made to alleviate food insecurity through indigenous foods it may reach the goal of zero hunger.The United Nation's Sustainable Development Goals (SDG) highlights the importance of food and nutrition security. The new South African National Policy on Food and Nutrition Security, and the Agricultural Research Council have been promoting the indigenous foods for food and nutrition security through research since 1994. Increased consumption of these indigenous vegetables will lead to greater demand, increased production and creation of markets and will thus enhance rural economies. Towards a better resilient food and nutrition system it should involve people as consumers and communities as central focus and also promote and preserve the utilisation of indigenous food.

The website http://www0.sun.ac.za/vivus has discussed on ways to decrease malnutrition in an agenda of 2030 and how to reduce the wastage of food. Freedman (2015) has studied the traditional food plant resources and the role of home gardens in food security). He has brought out the role of indigenous wild food plants into home gardens. How it can improve health and also income.

Home gardens are important in the economically developing world. In addition to providing immediate access to large number of vegetables grown organically it also provides a variety of foods throughout the year. This helps in a significant resource increasing the food security.Food security, however, is not exclusively a quantitative issue. It alo assures sustainable nutritional security.With the globalization of western food products over the past three or four decades, nutritionally-adequate traditional diets have been replaced by an increasing reliance on highly processed, refined, so-called fast convenience foods which have little or no nutritional value and has given rise to many noncommunicable diseases which include anemia, cancer, diabetes, goiter, obesity and vitamin A deficiency which are related to infant and childhood blindness. One of the most destructive health consequences of the world-wide loss of traditional diets is the decrease in regular consumption of fresh fruits and vegetables, which are a primary source of micronutrients (vitamins and minerals). Use of home gardens with biodiverse vegetables and fruits can rectify this problem, and will also help in the conservation of agrobiodiversity. Despite the erosion of traditional diets in many developing areas, some are following the traditional food: In Kholagaun village, Nepal, it is reported that 40% of all food eaten are from the wild sources and in addition to gathering wild food plants, cultivation in home gardens also plays an important role in household food and nutritional security. In South Africa, it has been found that traditional vegetables (both domesticated and wild) contributes to 31% of the value of all plants grown on residential plots. Despite being an ancient practice, transplanting wild food species has been reported in the literature. In addition to Nepal and South Africa, wild food plant cultivation has been documented in Guatemala, Kenya and also Thailand. Being a rich source of micronutrients, it creates income by the sale of surpluses in local and national markets. In Africa, the traditional african leafy vegetables are popularly used. Gomes et al., (2011) have studied the role of organic honey from Portugal.

Under global warming a lot of attention is given to the food-energy-water and adaptation to the climate. The Food and Agriculture Organization has a formal definition both for food and nutrition securities.Food security refers to having physical, social and economic access to sufficient, safe and nutritious food, which meets the dietary needs and food preferences for an active and healthy life. Nutrition security means access to adequate nutrition, its utilisation and absorption of nutrients in food to be able to live a healthy and active life. Both for food and nutrient securities it is important to have the availability, access and utilisation of these foods. The food-energy-water usually focuses on the resource management under climate and population pressure and governance. Food and nutrition security are better discussed at household levels. A correlation exists between household income, education level, local governance and food security. Overall food insecurity degrades nutrition security. In India an adequate food security has been obtained without the nutrition security. It has been seen that many multi-year droughts and undernourishment has brought about stunted and wasted growth especially in children who are below the age of six years. It depends on the house hold conditions, parental education, open defecation, fertility and healthcare conditions.India is one of the member in the World Health Assembly. However, the implementation of plans for nutrition targets have not reached the masses. Most of the times it has been found that the middle class consumes high calorie, low nutrition diet as like the developed countries. This is leading to an increase in obesity and diabetes. It is also the environmental degradation and undernourishment of local populations for malnutrition. It has been also noticed that land grab or any type of displacement has a great effect on nutrition security of small landholders. Many times it is the displacement and migration that affects the food security.Indian companies are a part in the global race to acquire fertile agricultural lands in many other countries.However, there are many internal displacements in India due to the loss of agricultural lands. There is a land protection act which will help the small landholders.The implementation of such regulations must take place in an efficient manner. It is found that the land displacement degrades parental education, household assets, food and nutrition security, children's nutrition and health. It is global warming that affects food and nutrition securities, it impacts on the food-energy-water due to extreme weather events resulting in crop damage. Deficiencies in micronutrients such as zinc and iron are quite visible. Exporting of food impacts food-energy-water along with food and nutrition security.Food and nutrition security involves the science to the household level. Guenther et al., (2008 b) studied agriculture, food and nutrition.

Bosco Bua and Christopher Onang (2017) have validated the role of African indigenous vegetables for food and nutrition security in Uganda. African indigenous vegetables (AIVs) play an important role in house hold food and nutrition security especially in Uganda. However, there is limited information on the value chain. Yet, the contribution of AIVs in terms of food and nutrition security is undisputable. A cross sectional survey was conducted, which included 60 respondents (farmers, key informants and extension officers) to assess the production system of the AIVs, profile an inventory and diversity of the cultivated AIVs. Semi-structured questionnaires and direct field observation were used to gather the necessary data. The findings revealed that AIVs form a very important part of the diets of the local people. Many constraints were reported as, lack of quality seed, competition from exotic vegetables, low market demand and pest and disease infestation.There is also a need for sensitization of farmers on the benefits of AIVs, developing an effective means of improved seed production and distribution to replace the farmer-to farmer seed exchange and conservation of AIVs germplasm as the strategies to improve AIVs production in the study area. According to Uusiku et al. (2010), consumption of AIVs helps in maintaining a balanced diet and averting the chronic effects of malnutrition. However, most of AIVs have been neglected or underutilized There is also a need for sensitization of farmers on the benefits of AIVs, to develop effective and improved seeds. Conservation of AIV's germplasm as the strategies to improve AIV's production in the study area. Cowpeas, hibiscus spp., pumpkins and crotolaria have been commonly grown.The reasons for cultivation of AIVs included food, medicine, nutritive value, resilience to adverse weather conditions and resistance to pests and disease.

The food system in South Africa is dichotomous which is characterized by a distinct, dominant agro-industrial and an informal food system. This hampers the development of smallholder producers. The major agro-industrial food systems have led to improvements in food supply but has also resulted in significant trade-offs with agro-biodiversity, dietary diversity, environmental sustainability and socio-economic stability especially amongst the rural poor. Therefore, South Africa needs to deliver on sustainable and healthy food systems under environmental change. Now it is important to bring the underutilized, indigenous and traditional crops into the food system, which can offer good opportunities for developing a sustainable and healthy food system. These will also help in achieving societal goals like employment creation, wellbeing and sustainability of the environmental.It can improve nutrition, income, restore biodiversity and pressure on local food culture and poverty alleviation. Accordingly, it is now well recognized that food security cannot be delinked from nutrition security to which consumption of indigenous vegetables significantly contributed (Keatinge et al., 2011). Consequently, continued investments in AIVs research and development are vital for battle against poverty and hunger alleviation worldwide. This is because the indigenous vegetables offer unique opportunities to diversify farming systems to ensure food security at household levels as they are nutritious and offer cheaper alternatives to the more expensive with lower nutritional value.

According to PROTA [Plant Resources of Tropical Africa (PROTA), 2005], Africa has diverse vegetable species out of which the primary used vegetables encompasses 53 different botanical families; of which more than 60% are pan-African. However, the majority of African vegetable species are of localised importance due to the limited distribution. The potential of indigenous vegetables towards food and nutrition security is unexploited. More knowledge and information are needed on the diversity of indigenous vegetables grown by the local communities. According to Vorster and Jansen van Rensburg (2005) there are a number of traditional vegetables which are used as a Source of food in South Africa. In a study done by Musinguzi et al.(2006) it has been found that many indigenous plants are being utilized in Uganda. Diouf et al. (2007) found that there are many production constraints.

The promotion and integration of indigenous vegetables into farming systems assist in their protracted use for food and nutrition security and conservation.

Also, indigenous vegetables can produce better yields in poor conditions with fewer inputs and less labour than other staples making it suitable for the households (Flyman and Afolayan, 2006). According to Tumwet et al. (2014) AIVs, provide essential nutrients and are also integral health boosters. They are the sources of traditional medicines.Many studies on chemical composition of indigenous vegetables and fruits have shown that they contain invaluable amounts of phytochemicals including crude protein, fat and oil, energy, vitamins and minerals.

In a review, written on the status of conservation of the indigenous leaf vegetables and fruits of Africa (Uusiku et al., 2010 and Adebooye and Opabote, 2004) it was found that possibly women are the ones responsible for the cultivation, collection and preparation of vegetables for food and nutrition security. In MSc thesis by Keller the views of the african farmers have been explained (Keller, 2004). Infact, High and Shackleton (2000) observed that home gardens are traditional land-use practices near the homestead where crop plants especially vegetables are grown for household consumption.Vorster et al., (2005) reported that the younger generations dislike vegetables because of the bitter taste, non-attractiveness and palatability(Vorster and Jansen van Rensburg, 2005). Additionally, this may also be attributed to disappearing knowledge on the production of AIVs. This is because traditionally information Sources were passed on from older generations to the younger generations through the oral culture. Abukutsa (2007) observed that extension services were either non-existent or inefficient. Accordingly, conventional agricultural extension workers normally discouraged farmers from intercropping or nurturing of AIVs in their fields because of the competition with the planted crop and thereby reducing the yields. In fact, lack of adequate knowledge on how to produce, store and distribute quality seeds may be another reason which is normally overlooked by the farmers among themselves. Masayi and Netondo (2012) found that the introduction and adoption of exotic vegetables may be responsible for the disappearance of some of the AIVs. Sensitization of the public would therefore, help to reorient the public demands towards such AIVs hence stirring up the need. Therefore, the protection of the habitat for the wild species and conservation of the forests with rich species and wetlands would be a valuable strategy because this would safeguard the continued existence of AIVs as well as preserve the germplasm which is very vital for future plant breeding. It is suggested that the government should multiply the seeds of those AIVs and distribute to the farmers for planting and slowly take them into the main stream.

Traditional Indian Foods

Indian cuisine consists of a wide variety of regional and traditional cuisines which are native to the Indian subcontinent. Given the range of diversity in soil type, climate, culture, ethnic groups and occupations, these cuisines vary substantially from each other as they utilize locally available spices, herbs, vegetables and fruits. Indian food is also heavily influenced by religion, in particular Hindu, cultural choices and traditions (Dias, 1996).

The various trade relations and colonialism have played a major role in bringing some foods to India. It brought a number of new vegetables and fruits to India.A number of these such as the potato, tomatoes, chillies, peanuts and Guava were brought in many regions of India (Balasubramanian, 2008) .

It is the Indian cuisine which made good international relations.It was the trade of spices between India and Europe which brought changes in cuisines across the world. It was also taken to the Middle East, North Africa, sub-Saharan Africa, Southeast Asia, the British Isles, Fiji and the Caribbean. The cuisine of India is 8, 000-year old.Under this various groups and cultures interact in the Indian subcontinent which gives rise to diverse flavours. Trading with Britain and Portugal has added a lot to the diverse Indian cuisine(Krishna Gopal Dubey, 2011;Achaya, 2003).Earlier it was mainly the consumption of legumes, vegetables, fruits, grains, dairy products and honey. However, staple foods eaten today include a variety of lentils (dal), whole-wheat flour (atta), rice and pearl millet (bājra) which has been cultivated in the Indian subcontinent since 6200 BCE(Achaya, 2003). During the Śrama a movement some people took to vegetarianism (Padmanabh, 2000, 2001). Many varieties of fruits, vegetables and grains are grown throughout the year.Later during the Yoga tradition a classification system in food came into existance and the food was categorized as saatvic, raajsic or taamsic (Paramahansa Yogananda, 1973, Maharishi Mahesh Yogi, 1990). Davidson (2014) has written a book on the Oxford companion of food. Davidson (2014) has written a book on the Oxford companion of food.

According to Bhat, Shripad (2012) a food system which has a long tradition would never be changed immediately. Change of place or career could be the reason for this to happen.Apart from these, political and historical reasons may also cause changes in a particular food system, thereby creating society's health and total food insecurity. We can identify some of the important reasons for the change in traditionally developed food sequence. Due to the use of fast foods the traditional foods have taken a toll. The fast foods are low in vitamins which were otherwise, naturally provided to the body through vegetables green grass, seed and spices. This upholds the concern over the changed food system in our society and its impact. There have been constant efforts to gain attention towards changed traditional food system and also the impact of health and social problems caused by it. A research study has been conducted by the National Botanical Research Institute on this subject.

Many ancient Indian recipes have been lost, however, the ancient texts give details of what was eaten in those times. It was in the Gupta dynasty that tea was introduced along with many new cooking methods. Later due to the influence of Central Asian cultures, a mix of Indian and Central Asian cuisine became popular. Most of the seasonings were done with saffron (Rosengarten Jr, Frederic, 1969).In the Medieval Indian Manuscript Nimmatnama-i-Nasiruddin-Shahi (circa 16th century) shows the popularity of samosas.

Traditional foods and beverages of different Indian states

Cereals and Millets based traditional foods

Staple foods of Indian cuisine include pearl millet (bājra), rice, whole-wheat flour (atta), and a variety of lentils, like masoor (most often red lentils), tuer (pigeon peas), urad (black gram) and moong (mung beans).Lentils may be used whole, dehusked like dhuli moong or dhuli urad. It can also be split.Split lentils or dal are used extensively(John et al. 2010). Pulses, like channa or cholae (chickpeas), rajma (kidney beans) and lobhiya (black-eyed peas) are very common, especially in parts of north India. Channa and moong are also processed into flour (besan).Indian dishes are usually cooked in oil from vegetables. Use of peanut oil is popular in northern and western India, mustard oil in eastern India. Shankar et al.(2005) have studied the response of serum lipid on introducing ghee as a partial replacement for mustard oil in the diet of healthy young Indians. Indian coconut oil along the western coast, especially in Kerala and parts of southern Tamil Nadu and also gingelly (sesame) oil are widely used in the south giving a fragrant and nutty aroma to the food items (Sharma, 2012). Gradually a flower revolution came into existence which include sunflower, safflower, cotton seed and soybean oils(Subodh Kapoor, 2002). Hydrogenated vegetable oil, known as vanaspati ghee, is another popular cooking medium which is very unhealthy due to saturated fats (Laura , 2009). Butter-based ghee or desi ghee, is used on/in many products.However, due to the fat content caution is needed while consumption. Many types of meat are used in Indian cooking, the most common being the chicken and mutton. Fish and beef consumption are prevalent in some parts of India, but they are not widely consumed except for coastal areas and the north east (fish and beef consumption are prevalent in some parts of India, but they are not widely consumed except for coastal areas and the north east. A Review has been written (Subramaniam and Vellingiri 2014) describing the outline of meat consumption in the Indian population.

Lentils

Lentils are a staple ingredient in Indian cuisine.The most important and frequently used spices and flavourings in Indian cuisine are whole or powdered chilli pepper (mirch, introduced by the Portuguese from Mexico in the 16thcentury) black mustard seed, sarso), cardamom (elaichi), cumin (jeera), turmeric (haldi), asafoetida (hing), ginger (adrak), coriander (dhania and garlic (lasoon) (Priti Chitnis Gress, 2008). One popular spice mix is garam masala, a powder that typically includes seven dried spices in a particular ratio, including black cardamom, cinnamon (dalchini), clove (laung), cumin (jeera), black peppercorns, coriander seeds and anise star(Devagi, 2007).Each culinary region has a distinctive garam masala blend. Goda masala is a similar spice mix but is sweet.This masala is widely used in Maharashtra. Some leaves commonly used for flavouring include bay leaves (tejpat), coriander leaves, fenugreek leaves and mint leaves.The use of curry leaves and roots for flavouring is typical of Gujarat (Cuisines of Andaman and Nicobar Islands, 2012) and also in South Indian cuisine (Man and Ellis, 1932). Sweet dishes are usually seasoned with cardamom, saffron, nutmeg and rose petal essences for aroma and nutritive value.In each state of India there exists a lot of variation in the cuisine.This is mainly due to the local culture, geographical location (proximity to sea, desert or mountains) and also the purchasing power and the local economics. Variation is also found seasonally, depending on which fruits and vegetables are ripe.

Traditional food in Indian States described in alphabetical order: Bhandari (2009a) has looked into the facts of Indian states at a glance.

Andaman and Nicobar Islands

Seafood plays a major role in the cuisine of the Andaman and Nicobar Islands. Staples of the diet of the Indigenous Andamanese traditionally include roots, honey, fruits, meat and fish, usually procured during hunting. Fish curry, chilli curry, macher jhol and prawn curry are famous from traditional times.

Andhra Pradesh

The food in Andhra Pradesh comes from two regions Rayalaseema and coastal Andhra. It uses a lot of spices and tamarind. Seafood is common and rice is the staple food. Many lentil preparations such as pappu (lentils) and pulusu (stew) and spicy vegetables are prepared. Green leafy vegetables like bottle-gourd and eggplant are used in the preparation of dal.Among pickles mango-based pickles like avakaya and maagaya, gongura (a pickle made from Kenaf leaves) are common.Traditionally the most common dishes with nutrition are chikkudukaya vepudu (sem) recipe, munakkada masala kura (drum stick), dondakaya menthi karam koora, gongura pachadi recipe - Andhra style sorrel leaves chutney, aratikaya vepudu recipe - raw banana, sarva pindi/gine pindi recipe (rice flour and chana dal flatbread, aratikaya masala pulusu recipe - raw banana stir fry, peanut chutney with tomatoes, masala akki roti with spinach recipe - sarvapindi, chama dumpa pulusu recipe which is made with arbi along with sesame tamarind sauce, pappu chekkalu recipe and spiced rice cracker, vankaya kothimeera karam with brinjals cooked along spices and coriander, vankaya kothimeera karam and oats pesarattu made out of oatmeal and green gram are commonly eaten.Usually, the Andhra food is very spicy. Guntur in Andhra pradesh is the largest producer of chillies in the state. They are known for their pungency and bright red colour. Fish and prawn are curried in sesame and coconut oils and flavoured with freshly ground pepper and eaten with rice.Traditionally, the Andhra meal is served on a clean banana leaf. The main course includes rice, dal, vegetables like yam, ladyfinger, brinjal and bitter gourd. The spicy mango and shrimp chutney's and pickles are traditionally eaten. In Chettinad area rice is the basic cereal. Peas, chicken, fish and meat are commonly eaten. Sun-dried and salted vegetables are also used. These traditional foods help in enhancing the nutirents in all seasons. Andhra Pradesh Cuisine, 2012.

Arunachal Pradesh

The staple food of Arunachal Pradesh is rice which is eaten with fish, meat and leafy vegetables(Cuisine, India cuisine. Incredible India, 2012).Traditionally the common dishes eaten are lettuce prepared by boiling with ginger, coriander, and green chillies (Edgar and Showick, 2010). Preparations like Kharzi, zan, khura, thenthuk, kajjikayalu, momos, khapse and bamboo shoots are very common. Popular snack is boiled rice cakes which are wrapped in leaves. Among the Monpa tribe, a dish known as thukpa is common.It is a kind of noodle soup (Assam General Knowledge, 2012). Looking into the tribal community meat, fish, eggs, beef, chicken, pork and mutton are widely eaten. Among the beverages, apong or rice beer is common which is prepared from fermented rice or millet (Rajatananda Das Gupta, 1982).Among pickles very common preparations are with usirikaya (gooseberry or amla), nimmakaya (lime) and tomato (Arunachal Pradesh food, 2013). Dahi (yogurt) is supplemented for tempering to reduce the spiciness.Breakfast items include dosa, pesarattu (mung bean dosa) vada and idli.The food is spicy and hottest of all Indian cuisine. The cuisine is influenced from Andhra.A lambstew made from lamb trotters seasoned with cassia buds, cardamom and potli ka masala (sandal wood, khas roots and dried roses) is common. This is eaten with sheermal, which is a kind of bread. The richly flavoured kacchi biryani is prepared with tender, marinated lamb and rice cooked together (amazing arunachal.com).

Assam

In Assam people traditionally eat khaar which is a meat delicacy. Other very common dishes are Duck meat curry, maasor tenga (tangy fish curry, squab, snails, silkworms, insects, goat, pork, venison, turtle, monitor lizards, aloo pitika, xaak aru bhaji (herbs and vegetables) and ou khatta. Vegetables are eaten in fresh, dried or fermented forms (Norman , 1980). Rice is the staple food and many varieties are prepared with rice. Fresh water fish is very common. The food items are prepared by either barbecue, steam or boiling. A traditional meal starts with a khar. Rice beer and rice wine are common (Calcutta Cuisine - Cuisine Of Kolkata India, 2019).Chewing betel nut with Paan is common (Food Habits - Banglapedia, 2019).

Bengal

Mainly fish, rice, milk, and sugar are the ingredients in Bengali cuisine (Majumdar , 2013).It has four types of dishes, charbya the food that is chewed like rice or fish; chosya, food that is sucked, like ambal and tak; lehya are foods that are meant to be licked, like chuttney and peya are the drinks (Bhandari, 2009). The fish varieties are usually perch, mullet, crab, carp, prawn, crayfish and lobster. Golden mustard oil is mostly used as a cooking medium. Spices are a vital part of all food preparations. Dimer Dhokkar Dalna and Luchi, a common Bengali dish of steamed cakes made with eggs, potatoes and spices and simmered in an aromatic curry, bhapaa aloo, is potatoes in panch phoron, coconut paste and mustard oil.Palatable prawns are cooked in hot spices and then bathed in the creamy flavor of coconut milk**,** Sorsebata Ilish Mach**,** 'Hilsa' fish is the highlight of the local cuisine. The fish is put in turmeric water and simmered in a mustard-poppy seed paste along with the five-spice mix. Bengali lamb curry**,** lamb put into yogurt for sometime and cooked in mustard oil along with mustard seeds, almonds and castor sugar. Doi machch is a comforting fish curry. Tender chunks of fish are cooked in the soothing flavor of yogurt along with mild spices.The use of chilli pepper is common as in the Chittagong district of Bangladesh (The Telegraph, 2016). Use of mustard oil along with large amounts of spices is common in most of the preparations.Usage of deep fried bread is also a common dish known as luchi.Fresh sweet water fish is prepared in many ways, like steaming, braising or stewing in vegetables and sauces based on coconut milk or mustard.Shondesh and rasgulla are popular sweet dishes made of sweetened, finely ground fresh cheese. The jaggery rasgullas are very famous. These recipes were originated in Bengal(Mohanty, 2017;Martin et al.2001;Bhandari 2009b). The government of west Bengal has acquired the GI status of rasgulla.

Bihar

The main Bihari cuisine is litti chokha which is a baked salted wheat-flour cake and is filled with sattu (baked chickpea; kabuli chana flour)along with some spices.This dish is served with baigan bharta(made of roasted eggplant (brinjal) and tomatoes) (Chhattisgarh Cuisine in India, 2005). Among meat dishes, meat saalan is a popular dish prepared from mutton or goat curry in which cubed potatoes are added with garam masala. Another popular dish in Bihar is Dalpuri.It is wheat-flour bread which is salted, filled with boiled, crushed and fried gram pulses. Malpua is a popular sweet dish of Bihar.It is prepared with a mixture of maida, milk, bananas, cashew nuts, peanuts, raisins, sugar, water and green cardamom and fried in oil. Chana Ghughni is a spicy-tangy evening snack item from the food of Bihar.Another notable sweet dish of Bihar is balushahi, which is prepared by a specially treated combination of maida and sugar along with ghee and the other worldwide famous sweet, khaja is also very popular which is made from wheat flour, vegetable fat and sugar. Laung-latika is a traditional dessert of the food of Bihar, Silav near Nalanda is famous for its production. Khajuria is the most commonly prepared snack of Bihari food in which a mixture of wheat flour and jaggery is deep fried. Thekua, is another sweet dish made of ghee, jaggery and whole-meal flour, flavoured with aniseed is very popular(Chhattisgarh delicacies, 2012).

Chhattisgarh

Chhattisgarh cuisine is very different from the rest of India.The staple crops are Rice, Bajra and Jawar. The tribal people eat mushrooms, bamboo pickle and bamboo vegetables (Kumar, 1994). Muthia is prepared with the rice batter which is seasoned with various spices. The dish is not fried but is steamed and is very healthy and nutritious.Aamat is the liquid gravy prepared using mixed vegetables which are cooked along with the ginger garlic paste and various spices. Bara is a snack made from the fermented Urad Dal.Various vegetables and spices are added along with the fermented lentils.It is a famous dish of the state.The requirement of minimum oil makes this food item light and healthy to eat.Faraa is another famous dish with main ingredient being rice.This is mixed with spices and are steamed.Dubki Kadi is a pulse-like dish which is cooked with curd. The pakodas of gram flour are added to the dish. They also brew liquor from the mahuwa flower palm wine(tadi in rural areas) (Chhattisgarh delicacies, 2012).

Dadra and Nagar Haveli

The staple food of the tribal includes rice and ragi (nagli). They also consume kodra and bajra. Ubadiyu is made of vegetables, beans and green leafy vegetables. The common foods include rice, roti, vegetables, river fish and crab. The staple food of the tribal includes rice and ragi (nagli). They also consume kodra and bajra.People drink buttermilk and eat chutnies made of different fruits and herbs (Pandey, 2018).

Daman and Diu

Seafoods are very common.Normally, rotli and tea are eaten in breakfast, rotla and saak (subzi) during lunch and chokha along with saak and curry are taken for dinner.Peas boiled in salt are very common. During festivals they prepare items like puri, lapsee, potaya, dudh-plag and dhakanu (Indian Street Food in Delhi, 2013).

Delhi

The most common foods are rajma-chawal, curried red kidney beans with steamed rice, paranthas, chaats, butter chicken, kebabs, chole bhature, biryani, nihariand vegetable and meat rolls.For the stuffed flatbread (paranthas with different vegetables) the Paranthewali Gali in Chandani Chowk is famous. As Delhi is cosmopolitan it has many different types of food traditions (Pat Chapman, 2009).The common street foods available are Kababs, kachauri, chaat, sweets, kulfi ice cream, sandwiches and patties (Cuisine of Delhi (2013). Tarla Dalal (2001) has described Achaar aur Parathe in an exotic manner (Cuisine of Delhi, 2013).

Goa

Xacuti is a Goan curry made with roasted grated coconut and coconut milk with the peri-peri chilli, the most important part of Goan spices. Rice is the staple food. They use coconut milk in plenty.Coconut is an essential ingredient of Goan Cuisine.Commonly eaten meats are Pork, Lamb and Chicken.Kokum is another important ingredient, it is sour, deep red colour fruit.Kokum drink and solkadi made out of coconut milk and kokum, aid digestion after meals.A pinch of salt, mustard seeds, cumin, garlic, asafoetida and red chillies give the typical flavour.Among the sweets are bibinica and dod bibinca made baked one over another and traditionally have 16 layers. These pancakes are made up of coconut milk, flour, sugar, egg and nutmeg. Dodol is jaggery flavoured fudge made from palm, jaggery, rice flour and coconut milk. Feni is a locally brewed wine made with cashew nut.The most popular curry dish in Goa is Pork vindaloo. Use of kokum is very distinctive in Goan food.The staple food is fish among the non vegetarians. Kingfish is very common and others are pomfret, shark, tuna and mackerel. These are usually eaten with coconut milk. They also eat commonly the shellfish, crabs, prawns, tiger prawns, lobster, squid and mussels (Goan cuisine, 2012).

Gujarat

khandvi is also known as Patuli or Dahivadi and made with gram flour.This rolled, bright snack is tempered with coconut, mustard seeds and curry leaves. Handa is a savory Gujarati dinner dish.It is mainly vegetarian. The thali food has many items which includes roti (rotlii in Gujarati), daal, kadhi, rice, sabzi/ shaak, papad and chaas (buttermilk).The sabzi is a dish of different combinations of vegetables and spices which may be stir fried, spicy or sweet(Aruna and Arlene, 2012).Kathiawari and Kutch, Southern Gujarat (Surat)Saurashtra Kathiawari and Kutch foods are very famous for their spicyness(Murdoch Books Pvt Limited, 2010).Undhiyu is a Gujarati vegetable curry is wholesome and very healthy.Made with a mélange of winter veggies like brinjals, surti papdi, potatoes, coconut, bananas and methi and is traditionally served in earthen pots. During the mango season, fresh mango pulp (keri no ras) is often an integral part of the meal. Use of spices varies as per the climate. Very popular dishes in khara snacks are sev khamani, khakhra, dal vada, methi na bhajiya, khaman and bhakharwadi.Farsan snack form is an essential part (Priya and Carol, 2005). Bardoli ki Khichdi is a fragrant rice preparation along with dal, spices, peas, potatoes and raw mangoes.People from Surat use gur (jaggery) in their cooking and this gives a sweet tangy flavour to their dishes. Surat has a plenty of green vegetables and fruits and used in food dishes.The use of jaggery in food items is common. The use of pulses is common. Asafoetida is used in all vaghars, as it helps in reducing the gas in the body. Colacasia leaves are a powerhouse of nutrients like Iron and when made into a delicious dish of Patra, it is tasty as well when a smooth paste of besan with spicy, sweet and sour flavours is applied over the leaves, rolled and steamed.

Haryana

Specific common dishes are kadhi, pakora, besan masala roti, bajra and aloo roti, churma, kheer, bathua raita, methi gajar, singri ki sabzi and tamatar chutney. In earlier days, the staple diet was bajra khichdi, rabdi, onion chutney, millet roti and bajra roti. Most common in non-veg food items is kukad kadhai and masala gravy chicken (Food of Haryana, Cuisine of Haryana, 2011). Mixed Dal is a staple diet in Haryana. It is a mixture of four to five dals which includes chana, toor, masoor and moong.Kachari ki chutney is made with potatoes and other vegetables mixed together. It has all types of ingredients which include garlic, onion, aniseed (saunf), turmeric, cumin seeds, red chillies and salt to taste. Among the beverages drinks like lassi, sharbat, nimbu pani and labsi(which is a mixture of bajra flour and lassi) are common (Kashmiri food, 2012).

Himachal Pradesh

Himachal pradesh specialities (2012) has discussed on the cuisine of the state. The traditional food include lentils, broth, rice, vegetables and bread. The delicacies include babru, pahadi, chicken, chamba style fried fish, chha gosht, kullu trout, gahat ka shorba, khatta, auriya kaddu, sidu, patande, chukh, rajmaha and til chutney(Kashmiri Wazwan, 2012). Udkiya Bhaath is a pahadi pulao which is a Source of many nutrients. It is a rice dish prepared with a variety of vegetables. Potatoes, ginger, garlic and onions are a necessary addition to the dish, along with yoghurt cooked with Indian spices along with onion, tomatoes, garlic, cinnamon and cardamom. Bhey is a truly unique dish, which is made from the stems of the lotus flower. Dhaam is a complete food that along with a great taste promises a healthy dose of nutrients.The dish includes dal, rajma, rice, curd, boor ki Kadi and is very well complemented with gur (jaggery). Aktori is prepared in the form of a cake or pancake made with the buckwheat leaves which is further cooked in the wheat flour.Kale chana ka khatta is another famous dish.

Jharkhand

Staple food in Jharkhand are rice, dal and vegetables.The popular dishes are Chirka roti, Pittha, Malpua, Dhuska, Arsa roti and Litti Chokha(High time to make most of mahua – Times of India.The Times of India, 2018).Rugra is an item which tastes like mushroom and is very delicious to eat.It is prepared with an indigenous traditional variety of vegetable, rich in protein and minerals and high on calorific value. This is thus a very healthy dish.Chilka Roti is prepared using rice flour and besan. Local alcoholic drinks include Handia a rice beer and Mahua daru, made from flowers of the "Mahua" tree (*Madhuca longifolia*) (TNN, 2009;Taste of Mangalore, 2002).Rice beer called as Handia is prepared by the local community.It is made from the ranu tablet which is a combination of 20-25 herbs, which are mixed with boiled rice and left to ferment. This drink gets ready within a week.Til Barfi is commonly prepared with sesam seeds.Meetha khaja is famous among the sweets.

Kashmir

In Kashmiri cuisine they add mutton, of which over 30 varieties are known(Anand, 2012). Rogan Josh is an aromatic lamb curry made with the fusion of various spices, yogurt, onions browned with little oil, which is a healthy low-fat dish.Wazwan is a multicourse meal in the Kashmiri tradition and is a delicacy. Birayani's are the speciality of Kashmir.The Jammu region is famous for its Sund Panjeeri, Patisa, Rajma (Kidney Beans)with rice and Kalari cheese.Dogri food includes ambal (sour pumpkin dish), khatta meat, Kulthein di dal, dal chawal, maa da madra and Uriya. Many types of pickles are commonly prepared that include mango, kasrod and girgle. Among the street foods various types of chaats, specially Gol Gappas, Gulgule, Chole Bhature, Rajma Kulcha and Dahi Bhalla are very famous. Matschgand which are minced meat balls are relished. Yakhni or Yoghurt Lamb Curry, Yakhni or Yoghurt Lamb Curry, Kashmiri Muji Gaad, Aab Gosht, Lyodur Tschaman, Momos, Thukpa, Skyu, Khambir, Paba.Traditional Kashmiri cooking is called Wazawan and comprises mostly non-vegetarian dishes. An important ingredient of Kashmiri Cuisine is curd and Asafetida (hing). Locally grown varieties of rice are sweetly fragrant and very light. The delicious saag is made from a thick-leafed green leafy vegetable called hak. Lotus stem, Morel mushrooms called gucchi, corn bread, fresh fish is a favourite.Smoked meat, dried fish and vegetables are stored and are used in winters. A special masala 'cake' is made from spice-blends, onions and locally grown chillies that can be stored for a long period of time and used in flavouring curries. Sauces are made from dairy-rich products.The fruits and nuts grown from the valley are used lavishly in the daily menus.

Karnataka

The staple vegetarian meal of Karnataka are bisi bele bath, jolada rotti, badanekai yennegai, holige, kadubu, chapati, idli-vada, ragirotti, akki rotti, saaru, huli, kootu, vangibath, khara bath, kesari bhath, sajjige, neer dosa, benne dose, mysore paak, haal bai, chiroti, ragi mudde and uppittu.Bisi bele bath is a delicacy in Karnataka which is made of rice, lentils, spices and vegetables. Idli, rava idli and Mysore masala dosa are very famous and were originated in Karnataka.The different cuisines include Mysore/Bangalore cuisine, North Karnataka cuisine, Udupi cuisine, Kodagu/Coorg cuisine and Karavali/ coastal cuisine. The Kodagu region is known for spicy pork curries while seafood is common in coastal Karnataka.The coastal districts of Dakshina Kannada and Udupi have slightly varying cuisines, which make extensive use of coconut in curries and frequently include seafood (Cuisine of Kerala, 2016).

Kerala Cuisine

Idiyappam (nool putto;is eaten with curry.Erissery (pumpkin and lentil stew), puttu and kadala curry, appam with ishtu (stew), ela sadya, parippu curry, dosa ghee roast with Kerala style sambar, idli sambar are other very famous dishes of Kerala.Yellow banana chips fried in coconut oil and lightly salted are liked by all. The rice appam, a pancake also called vellappam, is common to all Keralites and is eaten with a meat stew. Coconut milk is commonly used in all dishes (cuisine of Kerala, 2016). During Christmas wild duck with mappas are eaten. Meat porridge aleesa is widely eaten. Soups made from rice and wheat with added coconut / coconut milk and spices are taken before the meals. Among the desserts sweet known as muttamala containing a chain-like strings of egg yolk cooked in sugar syrup is relished.Many pachadies, pickels, chips and payasams based on milk, coconut milk, rice, dal and bananas are eaten.Rice is eaten with dishes like koottu, kalan and olan.Among desserts Kozhikodan halwva is made from maida (highly refined wheat) and comes in various flavours like banana, ghee or coconut.However, karutha halwa (black halwa) is made up of rice. (Bhandari, 2009c).India planning commission (2008) prepared a report on the development of Kerala.

Konkan

A hot and spicy digestive drink known as sol kadhi is a very famous dish which is made of coconut milk and kokam. The region of Ratnagiri is famous for its alphanso mangoes around the world. Coconut is an important ingredient of Konkan cuisine. For sourness people of this region use kokum (amsul) instead of tamarind. Among the fish, Bombay duck or bombil is eaten fresh or in dried form. The marinated duck fish is shallow fried. Bharli vangi and kadamb are like steamed idlis of Karnataka. Sprouted moong dal curry is prepared with cucumber, rice, jaggery and coconut.

Lakshadweep

The cuisine of Lakshadweep has mainly seafood and coconut.Spicy nonvegetarian, small and big sea-fish can be found in Lakshadweep, that are cooked exotically in coconut sauce. Drinking a lot of coconut water which is highly nutritious and is common.Coconut milk is the base for most of the curries.The dishes use Malabar spices.Among vegetarians, dosa, idlis and various rice dishes are common among the local people (Esteves and Lesley, 2008) .

Madhya Pradesh

Daal bafla, a popular dish in Madhya Pradesh, Rajasthan, and Gujarat. seekh kebab, which is a minced meat wrapped around a skewer (*seekh*) and then cooked on a coal fire is relished and is a traditional dish of Madhya pradesh. Palak puri is a common healthy and nutritious breakfast eaten along with raitha.In Gwalior and Indore milk is very common in food items (India's bountiful breakfast options, 2010). Meat and fish dishes as rogan josh, korma, keema, biryani, pulav and kebabs are commonly eaten.Paya soup, bun kabab and nalli-nihari are the dishes commonly eaten (Indian Food and the West New Emissary, 2012). Biryani pilaf is a common delicacy with mutton being the main type of meat. It is eaten with zarda which is a traditional South Asian sweet dish made by rolling rice in milk, sugar and is seasoned with raisins, pistachios, almonds, cardamoms and saffron. Dal bafla is quite common meal which consists of a steamed and grilled wheat cake dunked in rich ghee, which is eaten with daal and ladoos.In central Madhya Pradesh poha (flattened rice) is common and is usually eaten for breakfast along with jalebi. Beverages in the region include lassi, beer, rum and sugarcane juice.A local liquor is distilled from the flowers of the mahua. Date palm toddy is popular, however, in tribal regions, a drink from the sap of the sulfi tree is popular. On fermentation it produces alcohol. Among desserts mawa batti is famous in which the mawa dough is mixed with dried fruits and then rolled into balls, which are then deep fried and further soaked in sugar syrup and garnished with coconut powder.

Maharashtra

Bajra, wheat, rice, jowar, vegetables, lentils and fruits are the important ingredients of the Maharashtrian diet. The most popular dishes are pav bhaji, vada pav, puran poli, misal pav, modak, ragda pattice, bharli vangi, sreekhand, poha, sabudana khichdi, rassa, sol kadhi, pithl abhakri, aamti, Bhelpuri, Basundi, Kairi Cha Panha, Aamrus, Kadhi.For non vegetarians popular spicy meat dishes are the Kolhapuri sukka mutton, pandhra rassa, Savji mutton curries, tabmda rassa and bombil/Bombay duck. Kanda poha and aloo poha are eaten for breakfast. Shrikhand, is a sweet dish which is made from strained yogurt (Feed your desi mania at Nirula's, 2010). It is a main dessert of Maharashtrian cuisine (Flavours of Maharashtra at Renaissance, The Times of India, 2010). During summers, panna, a drink made from raw mango is commonly eaten (Lonely Planet blue list: the best in travel 2007; Hamlet, 2001). The cuisine of Maharashtra can be divided into two major sections; the coastal and the interior. In the Vidarbha region large amounts of groundnuts, poppy seeds, jaggery, wheat, jowar and bajra are used.A traditional meal consists of rice, roti, poli, bhakar along with varan and aamtee (lentils and spiced vegetables). Cooking is common with different types of oils. In Vidarbha savji food is popular which is very spicy and oily.

Malwani

The unique taste and flavor of this area is due to the use of Malwani masala, use of coconut and kokam. The staple foods are rice and fish. Various kinds of red and green fish, prawns, Mohari mutton, crab and shellfish curries (mashacha sar) are popular, along with kombadi (chicken) wade and mutton. There are many fish varieties available in the region.These are surmai, karali, bangada, bombil (Bombay duck), paplet (pompret), halwa, tarali, suandale, kolambi (prawns), tisari (shell fish), kalwa (stone fish) and kurli (crab). These are available in dried form, including prawns, called as sode. Local curries and chatani's are also prepared with dried fish. Different types of rice breads and pancakes are tandlachi bhakari, ghawane, amboli, patole, appe, tandalachi and shavai (rice noodles). These rice breads are usually eaten with coconut milk, fish curries and chicken / mutton curries.Sole kadi made from kokam and coconut milk is an appetizer. For vegetarians, alloochi bhaji, alloochi gathaya, kalaya watanyacha and sambara(black gram stew) are common.The sweets and desserts include ukadiche modak, Malawani khaje, khadakahde kundiche ladu, shegdanyache ladu, tandalchi kheer and tandalachi shavai ani ras (specially flavored with coconut milk).

Manipur

Chamthong or Kangshoi, is very popular in Manipur.It is a vegetable stew which consists of seasonal vegetables that are boiled and flavoured with sliced onions, cloves, salt, garlic, maroi and a bit of ginger.Eromba is a delightful combination of boiled vegetables and fish. Morok Metpa is a spicy and delicious chutney.Singju is a very nutritious Manipuri Salad which contains cabbage, coriander leaves, onions, ginger, lotus stems, the singju leaf and other available vegetables. The variations of dish also includes raw papaya, cabbage, banana flower and lotus root. The Singju is either flavoured with besan, powdered perilla seeds, chilli powder and other herbs and is highly nutritious. Paaknam is a savoury cake which is prepared from a thick batter of besan, herbs, vegetables and further flavoured with chilli and the traditional Ngari. It is made of a base of lentil flour stuffed with various ingredients such as banana inflorescence, mushrooms, fish and vegetables.The bake is later covered inside the turmeric leaves. The dishes are cooked like a vegetable stew and then flavored with either fermented fish called ngari, or dried and smoked fish. The manipuri salad dish is called as singju which is made with small pieces of cabbage, green papaya and other vegetables.It is garnished with local herbs and then toasted with sesame powder and lentil flour. It is served with boora which are fritters of various kinds, and also kanghou or oil fried spicy veggies. Cooked and fermented soybean is a popular condiment. The staple diet consists of rice, fish, large varieties of leafy vegetables (aquatic and terrestrial). Chak-hao Kheer is a delicious kheer.

Meghalaya

The common food items is jadoh which is made of red rice, cooked with generous amounts of pork meat. It is also cooked with chicken or fish. A mix of green chillies, onions, ginger, turmeric, black pepper and bay leaves is made. The pieces of pork are added and fried. Then the red rice is added. Doh Khlieh is a delicious salad made out of minced pork, onions and chillies. Nakham is a special kind of a dry fish, which is sun-dried or fire-dried. The fish is then fried and boiled in water, to make a thick, rich soup. Pumaloi essentially means powdered rice, which has been steamed. A unique pot called Khiew Ranei is used to cook the rice. Other common foods in Meghalaya include minil songa (steamed sticky rice), sakkin gata, Doh-neiiong is a pork curry dish and momo dumplings. Tungrymbai is a fermented soya beans, boiled and chopped pork, black sesame, ginger, onion and other spices. Fermentation of rice for the preparation of beer is common to be consumed and is known as Kyat (www.holidify.com).

Mizoram

A popular dish is Bai made from boiled vegetables (spinach, eggplant, beans and other leafy vegetables) with bekang fermented soya beans. Vawksa rep is smoked pork, chilies and local herbs. Saum is a fermented pork which is eaten with rice. Sawhchiar is made of rice and cooked with pork or chicken. HmarchaRawt is a very famous tangy and spicy chutney(Bhandari, 2009; Pran Nath Chopra , 1992).

Nagaland

Dried fish, prawns, ghost chili and preserved colocasia leaves are used commonly.It is known for exotic pork meats cooked with simple ingredients. Use of hot Bhut jolokia pepper, fermented bamboo shoots and akhuni or fermented soya beans are common. Hinkejvu is a simple dish made with colocasia, shredded cabbage leaves, mustard leaves and some French beans. The ingredients are boiled in water and the real flavour comes from the assorted greens. Akibiye is a simple dish made with colocasia and bamboo shoot.Fermented fish is known as ngari and Akini Chokibo is a snail meat which is included in pork dishes. Fresh herbs and other local greens are also widely used. The use of oil is less as the people prefer to ferment, dry/ smoke their meats and fish. Traditional homes in Nagaland have external kitchens that serve as smokehouses(Susie et al., 1993). In a meal rice, meat, chutney, stewed / steamed vegetable dishes are served flavored with ngari or akhuni.

Orissa

Machha Mahura is curry based dish prepared with fish and various vegetables. Pakhala is a rice dish prepared by simply adding water after the rice is cooked.It is then allowed to ferment overnight called as basi pakhala. Before fermentation, the dish is called saja Pakhala and is served along with onions, curd, green chillies. Dalma is a lentil dish prepared with lentils and vegetables which includes toor dal, green papaya, brinjal, pumpkin which are chopped and added. It is then garnished with spices and tempered with panch phutana. In Odisha use of local ingredients is common and spices are not used much.

Fish and other seafood, like crab and shrimp are popular, chicken and mutton are also consumed. For flavouring and seasoning the vegetables and dals Panch phutana, a mix of cumin, mustard, fennel, fenugreek and kalonji (nigella) are used widely.Chaatu Rai is one such traditional dish, where the mushrooms are chopped and a curry of mushroom with mustard and potatoes are prepared (Delish potpourri.2016).Garam masala and turmeric are commonly used for meat-based curries. Pakhala, is a fermented dish which is made of rice, water and dahi (yogurt). It is popular in rural areas especially in summer season. Popular Oriya dishes include Anna, Kanika, Dalma, Khata (Tamato and Oou), Dali (Different types of lentils like harada (red gram), muga (moong) and kolatha (horsegram).Saga (spinach and other green leaves) and alu-bharta(mashed potato) along with pakhala are popular dishes (lunch) in rural Odisha. Ou khatta is a fruit chutney which is part sweet part sour in taste. Among sweets rasagula, chhena poda, chhena gaja, chhena jhili and rasabali are very famous.

Pondicherry

Some favourite dishes include coconut curry, tandoori potato, soya dosa, podanlangkai, curried vegetables, stuffed cabbage , tandoori potato, soya dosa, assad, coconut curry, podanlangkai, curried vegetables, stuffed vegetables, palpayasam which is sweet milk and baked beans (Colleen, 2004). Poisson du jour is the dish which consists of grilled fish. Pate De Foie Gras has liver of the geese is the foremost ingredient in making this French dish. It is the mixture of ground meat that cooks well and minced fat which forms into the paste. Vegetables, spices are the additions in the dish. Bisque de Crevettes is made from shrimp. Cassoulet, which is a pork sausage is widely eaten (Indian cuisine: Puducherry, 2011).

Punjab

Punjab is known for tandoori (grilled) food. Tandoori chicken is very popular. Whole grains, vegetables and fruits form the major part of food items. Use of ghee, butter and cream is common. Use of whole wheat, rice and spices are common (Tarla Dalal, 2007). The regular home meal consists of roti, daal , dahi (yogurt), chutney and salad that usually has raw onion, tomato and cucumber. Masala Chai is relished as a drink. Many regional differences exist in the Punjabi cuisine due to different spice combination in the preparations [Tarla Dalal, 2007]. Most common dishes are makki di roti , sarson da saag and dal makhani; Punjabi cuisine, 2012). The masala in a recipe is prepared with onion, garlic, ginger, cumin, garam masala, salt, turmeric and tomatoes sauteed in mustard oil. Among the meat dishes Bhakra curry (Goat) and fish dishes are common (Clifford, 2005). Dahi, milk and milk derived products like lassi and paneer are widely used.Most common desserts are kheer, gajar ka halwa, sooji ka halwa, rasmalai, gulab jamun and jalebi.The desserts are mostly ghee or dairy-based where like almonds, walnuts, pistachios, cashews and raisins are used. Chhole Bhature and chhole kulche are also very famous. Use of lentils like black gram and yellow gram is common. The tandoor is used for cooking roties, paranthas, naan, kulcha and dishes like tandoori chicken and tandoori jhinga. Use of desi-ghee and white butter in dishes is common. Parantha, and roti are commonly prepared from the flour of wheat, corn, or other glutenous flour.Stuffed paratha with varoius vegetables is eaten along yoghurt, Rajma, the red kidney beans with pulao are relished (Subodh Kapoor, 2002). Among snacks gol gappa (panipuri) which consists of a round, hollow puri, fried crisp and filled with a mixture of flavoured water, boiled and cubed potatoes, onions with crushed bengal gram is liked by many.

Rajasthan

Grains like bajra, makkai, jawar and wheat are grown and eaten. Dried lentils and beans are also very popular. Juwar, bajra and makkai are used to make rotis and wheat is used to make Bati (dal-bati churma). Due to the scarcity of vegetables in the area gram flour is used widely (Game cuisine, 2014). Traditionally, the food consists of daal, baati, tarfini, raabdi, ghevar, bailgatte, panchkoota, chaavadi, laapsi, kadhi and boondi.Bhujia, mirchi bada, pyaaj kachori dal, kachori and mawa kachori are relished. Daal-baati is supplemented with choorma, a mixture of finely ground baked rotis with added sugar and ghee (Dinesh and Madalsa, 2014). Among non vegetarian dishes the common are laal maas, safed maas, khad khargosh and jungli maas(Care for some Sindhi savouries, 2011). A variety of gatte preparations are eaten since time immemorial like shahi gatte or masala gatte or gattes covered with spicy gravy. Gatte pulao is also common. gram flour is used to make gatte and pakodi. use of dry mango Powder known as amchoor and asafoetida. Among beverages, chaach and masala chai are common. Among raita using curds use of boondi is common. Among desserts ghevar, moong ki daal ka halwa, churma ladoo, aam ki launji, mawa kachori, badam ka halwa and gunjia are common.

Sikkim

Rice is the staple food and traditional fermented foods and beverages have constituted about 20 percent of the local diet.Use of finger millet, wheat, buckwheat, barley, vegetables, potatoes and soybeans are common. Dhindo, daal bhat, gundruk, momo gya thuk, ningro, phagshapa and sel roti are some of the local dishes. Momos are steamed buns with vegetable or meat fillings. Alcoholic drinks and thukpa, a noodle soup containing carrots, bell peppers, spinach, cauliflower and celery with spices is common(Lachu and Snortum,

2005). Sha Phaley has bread stuffed with ground beef and cabbage made into semi-circles and then deep-fried. Sinki soup made out of radish roots put into bamboo and pressed over with straw is highly nutritious.

Tamil Nadu

Use of rice, legumes and lentils along with blending of spices such as mustard, curryleaves, tamarind, coriander, ginger, garlic, chili, pepper, cinnamon, clove, cardamom, cumin, nutmeg, coconut and rose water (Sanjeev Dak Kapoor, 2012; Eugene and William, 1964).

The southern regions like Tirunelveli, Madurai, Paramakudi, Karaikudi, Chettinad and Kongu Nadu are known for spicy non vegetarian dishes(Dubey, 2011) and Chapman , 2009. Dosa, idli, pongal and Biryani are very popular dishes which are eaten with chutney and sambar. Preparation of orange peel pachari and cauliflower pakori have traditionally originated from Tamil nadu. Fish, chicken and goat meat are consumed. Among vegetarian food rice, vegetables and lentil preparations like rasam and sambar are commonly eaten.

Telangana

Telangana cuisine is typically spicy and consists of ingredients like tamarind, sesame seeds, red chillies and asafoetida both in vegetarian and non-vegetarian dishes. Roselle is a staple item used in making curries and pickles. Millet bread/roti is eaten commonly (Dum Pukht cooking, 2012).Both vegetarian and non-vegetarian foods are popular. Dishes like biryani, haleem, baghara baingan and kheema are common. Dahi (yogurt) is commonly added to meals (Krishna and Sukhdev, 1977). The very common dishes are Sarva pindi, malidalu sakinalu, garijalu, pachi pulusu, golichina mamsam, hyderabadi biryani, chegodilu, polelu, qubani ka meetha, thunti koora, potlakaya pulusu, rail palaram, bachali kura, puntikura chana dal.

Tripura

A famous dish known as Kosoi bwtwi contains beans, berma, onion and garlic with turmeric powder. A dish known as Mui Borok to its natives consists of dry and fermented fish known to the people as Berma.In Tripuri cuisine uses many vegetables, herbs, pork, chicken, mutton, fish, turtle, shrimps, crabs, freshwater mussels, periwinkles, edible freshwater snails and frogs are common(Dubey, 2011).Vaishnavite are usually vegetarians (Subodh Kapoor, 2002).Chauk is the traditional rice beer that is prepared by fermenting rice in water, Gudok which used to be prepared in young bamboo pipes, Kosoi Bwtwi is another Tripuri cuisine made of beans and Berma. Mosdeng serma is a chutney which is made of berma, garlic, red chilli and tomato.

Uttar Pradesh

Tehari is a vegetable pulao. The finest of all basmati rice is used to make this dish. Varieties of vegetables are cooked with the rice with different spices which provides it the taste and aroma. Among the vegetarian food; dal, roti, sabzi and rice are commonly eaten. Pooris , kachoris, chaat, samosa and pakora are the popular snacks(Royal Mughal Ladies and Their Contributions, retrieved 2015; The Sunday Tribune – Spectrum – Lead Article, 2003). Other common dishes include kebabs, dum biryani and various mutton recipes. Bedmi is the deep fried maida puri stuffed with urad dal and masala. The dum style of cooking on a slow fire is very famous(Mughlai cuisine tops popularity charts in capital, 2008).Dishes like kebabs, kormas, biryani, kaliya, nahari-kulchas, zarda, sheermal, roomali rotis and warqi parathas are the speciality.Kofta is a curry based dish with meat or vegetable made balls soaked in it. The cuisine uses the ingredients like mutton, paneer and spices like cardamom and saffron (Dewan et al., 2001). The tastes are from mild to spicy and is associated with a distinctive aroma and the taste of ground and whole spices (Dalal, 2010). Sheer qorma, ghevar, gulab jamun, kheer and rasmalai , white til is mixed in cardamom flavored sugar or jaggery syrup are some of the popular desserts in this region.

Uttarakhand

Kafuli is a popular dish from Uttarakhand is made with many green vegetables like spinach and fenugreek. It is healthy and wholesome and is a high protein diet that makes heavy use of pulses and vegetables. In the traditional way, it is cooked on wood or charcoal fire in iron utensils.Condiments like jeera, haldi and rai are used.Some exotic condiments like jambu, timmer, ghandhraini and bhangira are also used.Preparations unique to Uttarakhand tradition are rus, chudkani, dubuk, chadanji, jholi and kapa. kafuli, bhang ki chutney, garhwal ka fannah, phaanu, baadi, kandalee ka saag, chainsoo, kumaoni raita.Among dressed salads and sauces, kheere ka raita, nimbu mooli, mooli ka raita, daarim ki khatai and aam ka fajitha are common.The cuisine is from the two sub regions, Garhwal and Kumaon in which ghee is used a lot along with lentils /pulses, vegetables and rice (Buisness world, 2020).Use of Badi (sun-dried Urad Dal balls) and Mungodi (sun-dried Moong Dal balls) as substitutes for vegetables is very common. Among desserts people eat pua, singal, singodi, bal-mithai and malai laddu.

Common Non-alcoholic beverages in India

Tea is a staple beverage throughout India, since the country is one of the largest producers of tea in the world. The most popular varieties of tea grown in India include Assam tea, Darjeeling tea and Nilgiri tea. It is prepared by boiling the tea leaves in a mix of water, milk, and spices such as cardamom, cloves, cinnamon, and ginger. In India, tea is often enjoyed with snacks like biscuits and pakoda. Coffee is another popular beverage, but more popular in South India. Coffee is also cultivated in some parts of India. There are two varieties of coffee popular in India, which include Indian filter coffee and instant coffee.Lassi is a traditional dahi (yogurt)-based drink in India. It is made by blending yogurt with water or milk and spices. Salted lassi is more common in villages of Punjab and in Porbandar, Gujarat. Traditional lassi is sometimes flavoured with ground roasted cumin. Lassi can also be flavoured with ingredients such as sugar, rose water, mango, lemon, strawberry, and saffron (Lindsay and Amelia, 2008).

Sharbat is a sweet cold beverage prepared from fruits or flower petals. It can be served in concentrated form and eaten with a spoon, or diluted with water to create a drink.

Popular sharbats are made from plants such as rose, sandalwood, bel, gurhal (*hibiscus*), lemon, orange, pineapple, sarasaparilla and falsa (*Grewia asiatica*). In Ayurveda, sharbats are believed to hold medicinal value (WHO, Ranjit and Uton, 2001).Other beverages include nimbu pani (lemonade), chaas, badam doodh (almond milk with nuts and cardamom) and coconut water. Cold drinks unique to southern India include beverages, such as panner soda or gholi soda, which is a mixture of carbonated water, rose water, rose milk and sugar. Narenga Soda, a mixture of carbonated water, salt and lemon juice and soda nannari sharbat, a mixture of sarasaparilla Sharbat with carbonated water are most popular non alcoholic beverages in Kerala and Tamil Nadu. Street shops in Central Kerala and Madurai region of Tamil Nadu are most popular for these drinks which are also called 'Kulukki Sharbats' in Kerala. Compared to natural fruit juices, synthetic drinks do not contain nutrients unless they are fortified. Generally, synthetic drinks are prepared using preservatives, artificial colours and flavours such as cola, orange, mango and lime, and mostly they are carbonated. Carbonated beverages contain phosphoric acid and may damage the enamel of teeth, and affect appetite if taken in excessive amounts. Water used for preparation of beverages should be free from disease-causing agents and harmful chemical impurities. Beverages like buttermilk, lassi, fruit juices and coconut water are better alternatives to synthetic drinks. Tea and coffee are popular beverages. They are known to relieve mental and muscular fatigue. This characteristic stimulating effect is due to their caffeine content. A cup (150 ml) of brewed coffee contains 80-120 mg of caffeine and instant coffee 50-65 mg, while tea contains 30-65 mg of caffeine. Caffeine stimulates the central nervous system and induces physiological dependence. Generally, low doses (20-200 mg) of caffeine produce mild positive effects like a feeling of well-being, alertness and being energetic. Higher doses (>200 mg) can produce negative effects like nervousness and anxiety, especially in people who do not usually consume caffeine-containing beverages. Therefore, moderation in tea and coffee consumption is advised so that caffeine intake does not exceed the tolerable limits. Tannin is also present in tea and coffee and is known to interfere with iron absorption. Hence, tea and coffee should be avoided at least for one hour before and after meals. Besides caffeine, tea contains theobromine and theophylline. These are known to relax coronary arteries and thereby promote circulation. Tea also contains flavonoids and other antioxidant polyphenols, which are known to reduce the risk for coronary heart disease and stomach cancer. However, as a result of its caffeine content, excess tea consumption is deleterious to health. Decaffeinated coffee and tea are being marketed to obviate the adverse effects of caffeine.Coffee consumption is known to increase blood pressure and cause abnormalities in heartbeat. In addition, an association between coffee consumption and elevated levels of total and LDL cholesterol (bad cholesterol), triglycerides and heart disease has been demonstrated. Therefore, individuals with heart disease need to restrict coffee consumption. Also, those who experience adverse effects from caffeine should stop drinking coffee. Tender coconut water is a nutritious beverage. It has a caloric value of 17.4 per 100 gm. The concentration of sugar steadily increases from 1.5% to about 5.5% in the early months of maturation and this slowly falls to about 2% at the stage of full maturity. It contains most of the minerals such as potassium (290 mg%), Sodium (42 mg%), Calcium (44 mg%), magnesium (10 mg%), Phosphorus (9.2 mg%), iron (106 mg%), and copper (26 mg%). It is a oral rehydration medium and keeps the body cool. However, in patients with hyper kalaemia such as renal failure, acute adrenal insufficiency and in patients with low urine output, tender coconut water should be avoided.

Among the alcoholic beverages common in India are Beer

Most beers in India are either lagers (4.8 percent alcohol) or strong lagers (8.9 percent). Bastar beer which is prepared from Sulfi is very common. The Indian beer industry has witnessed steady growth of 10–17 percent per year over the last ten years. Production exceeded 170 million cases during the 2008–2009 financial year(Ravi, 2010). With the average age of the population decreasing and income levels on the rise, the popularity of beer in the country continues to increase.Other popular alcoholic drinks in India include fenny, a Goan liquor made from either coconut or the juice of the cashew apple.The state of Goa has registered for a geographical indicator to allow its fenny distilleries to claim exclusive rights to production of liquor under the name fenny (www.casemine.com). Hadia is a rice beer, created by mixing herbs with boiled rice and leaving the mixture to ferment for around a week. It is served cold and is less alcoholic than other Indian liquors. Chuak is a similar drink from Tripura. Palm wine, locally known as Neera, is a sap extracted from inflorescences of various species of toddy palms (Chandrasekhar, 2012). Chhaang is consumed by the people of Sikkim and the Darjeeling Himalayan hill region of West Bengal. It is drunk cold or at room temperature in summer, and often hot during cold weather. Chhaang is similar to traditional beer, brewed from barley, millet, or rice (Sarada, 2017; Basappa et al, 2009). RenuAgrawal and Basappa (1994) have shown that the traditional fermentation of paddy malt mash (containing 18.1% w/v dextrose equivalent) to paddy arrack using paddy husk as Source of inoculum yielded very low level of ethanol (4.25% v/v). However, when yeast isolates obtained from paddy husk as well as a potent ethanol producer like *Zymomonas mobilis* ZM4 and their combinations were used in the fermentation revealed that a combination of an yeast isolate PH 03 (*Saccharomyces cerevisiae*) and *Z. mobilis* ZM4 produced synergistically and statistically more ethanol (10.1% v/v) than the individual and other combination of cultures. Kallu (Chetthu Kallu) is a popular natural alcohol extracted from coconut and pine trees in Kerala. It is sold in local Kallu shops and is consumed with fried fish and chicken. Its alcoholic content is increased by addition of alcoholic additives.

Traditional foods around the Globe

Alan (2014) has discussed on the oxford companion of food.

Traditional foods of Malaysia

Malaysian Indian cuisine, are seasoned with curry leaves, spices, ghee and fresh coconut are added.(Why are banana leaf meals popular in Malaysia; asianinspirations.com.au. Retrieved 23 March 2018).Rice is served in the center and other dishes like pickle, vegetable, papadam (made of lentils) About food in Malaysia (2012). (http://blog.tourism.gov.my/banana-leaf-rice/). Nasi lemak is the national dish which consists of coconut rice, prawn sambal, fried anchovies, peanuts, cucumber slices and ayam rendang. Ikan baka, nasi kandar, roti canai, curry laksa, char kuay teow, mamak dish and hokkien mee are eaten (Nathan, 2013). In a dish called as nasi kandar briyani rice is served with curry made out of chicken, fish, beef or mutton. It is accompanied with pickled vegetables and papadum.Appam is a bowl-shaped pancake prepared with fermented rice batter and coconut milk.Attukal paya is a mutton leg soup locally called as Sup kambing.Laksha are rice noodles in tamarind broth peppered with mackerel and vegetables. Bak kut teh, is a pork rib stew with a salty broth. It is pan-fried shredded fish mixed with grated coconut and spices (Samuel, 2013). Kaya is jam made from coconut milk with the consistency of soft butter.When not used as a bread spread, it is jammed into a bun and steamed to perfection for breakfast. Pisang goreng is a deep-fried banana.Mamak Rojak is a salad made up of cucumber (shredded), potatoes, beancurd, turnip, bean sprouts, prawn fritters, spicy fried crab, fried octopus or other seafoods. Pasembur is a variant of Mamak Rojak common in northern peninsular Malaysia. It is especially associated with Penang.Mee goreng is made with thin yellow noodles fried with garlic, onion or shallots, fried prawn, chicken, chili, tofu, vegetables, tomatoes and eggs.Maggi goreng is a variant of mee goreng using Maggi instant noodles.Mee goreng mamak is made with spices, tomato sauce, potatoes and sweet soy sauce.Mee rebus is a dish prepared with egg noodles in a sweet and spicy sauce, green chillies and a hard boiled egg (Tan, Su-Lyn, 2018). Murtabak is a dish of savoury stuffed roti along with minced mutton, garlic, onion and folded with an omelette.In the vegetarian nasi lemak dried anchovies are substituted with vegetarian mock anchovie.Nasi Kandar is a steamed rice either plain or mildly flavoured. It is eaten with a variety of curries and side dishes.Parattal or Peratal is prepared in a traditional south indian style which is a dry curry cooked with meat or vegetables(Samuel, 2013).Among the snacks kacang putih is made up of flour, nuts or legumes and many types of spices roasted or fried.Achu muruku is also known as rose cookies.Banana chips are either deep-fried or slices are dried from bananas.They can be covered with sugar or honey and have a sweet taste, or they can be fried in oil and spices and have a salty or spicy taste. Another snack mix which consists of a variable mixture of spicy dried ingredients like fried lentils, peanuts, chickpea flour ghatia, corn, vegetable oil, chickpeas, flaked rice, fried onion and curry leaves. Kacang Kuda Rebus is a preparation of chickpeas in curry leaves, chillies and mustard seeds. Kerepek ubi kayu, and samosa are fried savoury filled with potatoes, onions, peas, lentils, macaroni, noodles or minced meat (Borneo, 2008). Among the desserts Adhirasam is famous. Ais kacang is a traditional ice machine which is used to churn out the shaved ice used in the dessert.Cendol are smooth green rice noodles in chilled coconut milk and gula Melaka. Sea food curry (2018) has given delicious recipes of malaysian Indian style.

Teh tarik is a traditional drink in which tea is sweetened using condensed milk. kopi tarik or pulled coffee instead of tea; teh halia in which the tea is brewed with ginger and with or without the tarik treatment and teh madras prepared with three separate layers of milk at the bottom, black tea in the middle and foam at the top. Among alcoholic beverages toddy or kallu is an alcoholic beverage which is commonly obtained from the sap of various species of palm tree such as the palmyra and coconut palms.

Traditional foods of Singapore

Chilli crab is the traditional dish of Singapore. Hainanese chicken is a rice dish with juicy steamed white chicken cut into thin pieces. The chicken is served over fragrant rice with light soy sauce. The dish is topped with cilantro and sesame oil and accompanied with a garlic-chili sauce. The rice is cooked in the chicken broth with ginger and pandan leaves giving it an unique fragrance. Nasi lemak otak, bak kut teh is a pork rib soup and kaya toast are all famous dishes of Singapore.

Traditional food of Indonesia

Satay is a national dish of meat skewers that are cooked over coals and served with rice cakes (ketupat) along with peanut sauce poured all over the satay. Beef Rendang is a dish originated from Padang, Sumatra. Nasi Rawon is a dish made of beef stew from East Java. It has a nutty flavour and has a deep, black color due to the use of keluak nut. Oxtail soup is usually fried or barbecued and has a soup base. Pempek or empek-empek is made of fish and tapioca. Siomay contains steamed fish dumplings and is served with steamed potato, cabbage, egg and served with peanut sauce. Among the desserts sweet martabak is common. Thalassery (2012) has described biriyani style of Singapore cuisine.

Traditional food of USA

Lance Richardson (www.traveller.com.au)has described the important traditional dishes of USA as under:

Tater tots are grated potato mini-balls which are cylindrical in shape and fried with a crispy exterior.

Hot dog is usually on a poppy seed bun with different fillings.

Apple Pie has a combination of sugar, buttery pastry and tart sliced apples.

Barbecue ribs

Barbecue food is relished by all americans.Reuben sandwich contains corned beef, Swiss cheese and sauerkraut. It is similar to American delis. Biscuits and gravy are traditionally made with butter or lard and buttermilk; the country gravy with meat drippings and (usually) chunks of fresh pork sausage and black pepper.

It contains ground meat and seasonings, made into a loaf shape using a loaf pan or hand-shaped, roasted which is topped with sauce or ketchup.

Grits is made from coarsely ground corn kernelsand is usually eaten with shrimp.

The typical Hamburger is made from five-meat blend and cooked in a cast iron grill. Usually with bacon, sliders with green chilies and juicy lucy. Clam Chowder is a fragrant soup prepared with the quahog shellfish with tender potatoes, salted pork, heavy cream and herbs.

It contains bagel, sliced and toasted, cream cheese, lox or smoked salmon, capers, cucumber slices, red onion sprouts with salt and pepper.

Deep-dish pizza is prepared in which the pan is baked.It gives the pizza its characteristically ample space for large amounts of cheese and a chunky tomato sauce.

Drop biscuit and sausage gravy

Drop biscuits and Sausage gravy is a flaky scone made with lard and buttermilk. People working on horse ranches, eat biscuits for breakfast which are smothered in a thick white gravy that is studded with bits of sausage.Texas barbeque is a corn milled into a rough powder and then boiled along with butter or bacon oil. Geechie boy grits with shrimp and andouille sausage are famous. It can be eaten with brussels sprouts and sweetened ice tea.

Tacos is very common in Los Angeles like the greasy nachos, which are michoacan-style goat stews. Fajitas mole verde, flautas and fried crisp taquitos stuffed with chicken are relished. Traditionally the Mexican dish consists of a small hand-sized corn or wheat tortilla topped with a filling. The tortilla is then folded around the filling and eaten by hand.Chicken taquitos served on a bed of lettuce topped with sauce, salsa and cheese. During the thanksgiving the most common preparations eaten on this day are turkey, cranberry sauce, pumpkin pie, green bean casserole and bellyaches. Sandwich bread, butter Cheeses with bread are common.

Gouda and Swiss are common cheeses in America. Other common cheeses in America are bergenost, cheddar cheese (orange cheddar is used the most within Americans), colby (as well as the blend colby-jack), cottage cheese, cream cheese, monterey jack cheese, m ozzarella, Provel, Swiss cheese and Velveeta (brand name of a common processed cheese). Among the rice dishes, Gumbo dish has originated in southern Louisiana during the 18th century. Calas, charleston red rice, chicken bog, dirty rice, glorified rice, gumbo, hawaiian haystack, hoppin' john, jambalaya rice and gravy, shrimp creole, spanish rice and yeung chow fried rice.

Traditional American desserts

Chocolate chip cookies, cheese cake, banana foster, peacon pie, whoopie pie, carrot cake, flavoured ice creams, brownie and apple cinnamon bread pudding. Boston cream pie is a very famous dessert.

Traditional foods of Europe in alphabetical order

Austria

Strudel is made of crisp dough with delicate layers of different kinds of fillings. Apple is mixed with raisins. Traditional Austrian cuisine is similar to that of Viennese cuisine. The dishes are wiener schnitzel, apfelstrudel (apple strudel) / sachertorte are internationally well-known and are regarded as typical Austrian dishes. It is influenced by many nationalities and cultures in the Austrian-Hungarian monarchy. However, in the north it has been influenced by the czech cuisine and the Hungary in the east.The south is influenced by Italy and Slovenia.These are mainly meat based containing either pork or beef. During Christmas, carp is traditionally served. Among the vegetables green leafy and root vegetables, different types of beans and pumpkins are common.Traditionally, maize and rye are very popular.Desserts and pastries like apfelstrudel, mohnkuchen (poppy seed cake) and sachertorte are common. Vegetable Soup/Gemüsesuppe, made up of typical traditional vegetables like cauliflower, brussels sprout, carrots, celery, leek and green beans is common. Viennese Schnitzel/Wiener Schnitzel had originated from the Italian cotoletta in which the crust is made up of flour, whisked eggs and bread crumbs.Cabbage noodles/Krautfleckerl are based on noodles and cabbage in which bacon and onions are added. For a typical flavour white wine and caraway are added. In the southern part, Potato dumplings/Erdäpfelknödel are famous. They are made of plain dough based on flour, breadcrumbs or potatoes. Apple strudel/ Apfelstrudel is another famous recipe in which the apples, sweet bread crumbs and raisins are mixed. Rum and cinnamon are added for flavour This mix is filled into a strudel pastry jacket which is made out of wheat flour.

Belgium cuisine

Belgian waffle is often massive and drowning in syrup with whipped cream and sliced berries on top. Since the middle age traditional Belgian cuisine is being consumed by people (Pappas 2008a). Some people also have an influence of the French cuisine (Jacobs and Fraikin, 2005). The common traditional foods in the area are; Potage au lait battu (milk soup), Hochepot (vegetable and meat stew) and Bifteck aux pommes de terres frites made out of beef (Pappas 2008a). Common food from Belgium are fries and are called Frieten in Flemish and Frites in French. Belgium is also famous for its variety and quality of beers, their fine chocolates, their waffles, mussels and the Belgian endive, which is also known as chicory. The combination of bitter, sour and sweet in one dish is characteristic of the Belgian cuisine. Other common traditional dishes include konijn met pruimen en geuze/ lapin à la gueuze (rabbit stewed in naturally fermented beer from the Brussels region), stoemp is a dish based on mashed potatoes and vegetables which is served with sausages), salade liégeoise/ luikse salade (salad with bacon, potatoes, french beans, onions and vinegar), Hespenrolletjes met witloof/ chicon au jambon (braised chicory, wrapped in ham in a cheese sauce topped with grilled grated cheese) and mosselen met frieten/moules frites (mussels and french fries). There are many other well known traditional dishes which include waterzooi (soup/stew made of chicken, potatoes and vegetables), vlaamse stoofkarbonaden/carbonades à la flamande (stew made of beef and onions).Another traditional food is paling in het groen/anguilles au vert (fried eel in a sauce made of green vegetables and herbs), traditional shrimp croquettes are made with brown north sea shrimp. The croquettes are served with a lemon wedge and fried parsley.Flemish stew/vlaamse stoofkarbonaden vlaamse stoofkarbonaden or Carbonades à la flamande is a recipe that dates back to medieval times. This sweet-and-sour stew is made with an old flemish brown beer. This beer is stored in oak barrels and is known for its sour taste. The main ingredients are beef and onions.Meat loaf, meat balls/vleesbrood, vleesballetjes are served with sour cherries.The sour cherry sauce is typical for Belgium. Main ingredients of this dish are minced meat (either beef/veal or beef/pork), eggs, breadcrumbs and onions. The sour cherries are poached in sugar syrup. The meat mixture is sometimes baked as a loaf and cut into slices before serving. Art of waffle making is very common in Belgium. Every household has a waffle iron in the home. The main ingredients of this recipe are flour, eggs, milk, water, butter/ margarine, sugar and yeast.

Bulgaria

Soup is a tarator, yogurt-base soup which is made of cucumbers, garlic, dill and sometimes walnuts. Under the traditional Bulgarian cuisine there is an influence of proto-bulgarian, thracian, slavic, turkish, greek and middle east. There is a blend of european, oriental and mediterranean cuisines. The Turks brought many new fruits and other varieties of existing local crops to Bulgaria (Gavrilova, 2005). Salad is served with every meal and cold and hot soups are relished by the people. Bulgaria has a diversity of dairy products, including Kiselo mliako (a yogurt), white brined cheese and also a variety of wines and other local alcoholic drinks, known as rakia, mastika and menta. They also include bread and pies, pulses (legumes), fresh and pickled vegetables, salads, soups, stews, casseroles, stuffed vegetables, kebabs, spicy sausages (pastarma, lukanka) and cheese dishes. Meat (pork, chicken, lamb and beef), fish and vegetarian dishes are served with staple foods such as rice or bulgur wheat. Typical dishes include banitsa (bulgarian pastry), lyutenitsa (pepper relish), sladkish ot tikva (pumpkin pie), tarator (cold soup), hotchpotch (vegetable and meat stew), stuffed cabbage/vine leaves and moussaka. A popular traditional dessert is Baklava. Cold soup tarator is a starter with main ingredients like cucumber, yogurt and walnuts. Garlic and dill are used for seasoning. Sometimes, in place of cucumber other vegetables like marrow, celery, lettuce or carrots are used. Veal stew/Teleshko 'Popska' yahnia, is a meat stew which is cooked with veal, mutton, beef, poultry or rabbit. The basis of Teleshko 'Popska' yahnia includes shallots and fatty meat with spices. Nettle with rice/ Kopriva s oriz, has been traditionally a common food. Consumption of leafy vegetables like nettles, dock, spinach and sorrel is very common. Nettle's are used throughout the country. These are cooked in different procedures in different regions. The combination of nettle with rice is a typical seasonal vegetarian dish. Pepper relish/Lyutenitsa; is a mash (sauce) of vegetables which is made of tomatoes, peppers and spices. Some newer recipes also add aubergines, carrots, potatoes and/or onions. Pepper relish can be served as an appetiser or a garnish. In the past it was often combined with leeks, white beans or cheese and was eaten as a main dish in the Bulgarian menu. Typical national desserts are pumpkin pastry/sladkish ot tikva. It is prepared with pumpkin in which dried fruits are added.

Croatia

Black Risotto is a mix of cuttlefish or squid, olive oil, garlic, red wine and squid ink.

Cyprus

Halloumi Çheese originated in 16th century has a mixture of goat and sheep milk which gives it a different taste and texture, making it ideal for grilling rather than melting.

Czech Republic

These bread 'Dumplings', known as smørrebrød resemble the bread loaves and are made from boiled flour, yeast, egg, salt and milk and eaten dipping in goulash.Semolina pastry and yogurt and pies are other common desserts (.

Denmark

Smørrebrød are open-faced sandwiches which are combined with buttered rye bread along with eggs, cheese, meat and/or fish. These are topped with either roast beef, horseradish, shrimp and red caviar.The cuisine of Denmark is comparable to that of other Scandinavian countries. Mostly people have survived on local and home grown food without a refrigerator. Important preservation methods have been salting, pickling and drying. Rye, barley, dried peas and salted pork are common.Pickled herring and cured dried fish have been the basis of the traditional cuisine. During the winters kale, cabbage, carrots and other root vegetables and potatoes are grown and utilized in food items.This increased the retail trade and expanded the fresh food market (Boyhus, 2005). In the late 19th century and at the beginning of the 20th century, roast meats with potato gravy and consommé with flour and meat dumplings were very common. Also beef boiled in a sweet and sour horseradish sauce with currants has been relished. Different meats like goose, pork or duck were roasted with apples and prunes, accompanied by sweet and sour red cabbage and caramelised potatoes. Vegetables were often served in a 'white sauce'. Minced meat was also very popular during that period. Old preservation methods like curing or cold-smoking for fish have been replaced by hot-smoking and smoked fish. Sweet soups, like apple soup or other fruit soups are common in danish cuisine (Boyhus, 2005). Denmark is one of the world's largest exporters of pork and pork products. Therefore, it has been observed that Leverpostej (liver paste) and Spegepølse (fermented sausage, salami) are staple foods in most of the danish households. Cheese production is extensive and comprises of traditional products like Rygeost (smoked fresh cheese with caraway seeds).Danish pastries (known as wienerbrød), smørrebrød (open sandwiches) and frikadeller (fried meatballs) are common. Patty shells with chicken and asparagus/tarteletter med høns i asparges.These are small pots made with wheat flour. These are baked and filled with a thick stew of asparagus and chicken. Hamburger steak/Hakkebøf is common which are made from minced beef, breaded in the flour and then fried. It is served with potatoes and topped with gravy and fried onions. In Fried plaice, pieces of plaice are breaded in egg and breadcrumbs which are pan-fried. They are usually served with potatoes and browned butter, parsley butter, parsley sauce, salsa verte or remoulade (similar to tartar sauce). Traditionally, cold or hot potato salad, cucumber salad or gooseberry stew are eaten along with it.Strawberry stew with cream is made of strawberries boiled in water after which sugar and a thickener are added. It is served with milk or cream.Vanilla can be used for flavouring. Apple charlotte is a famous danish dessert in which apples are boiled and mashed and breadcrumbs and sugar are fried in butter. A layer of apples is then topped by a layer of the breadcrumb mix. Whipped cream and fruit jelly are used to garnish.

Estonia

Rye bread is obtained by natural fermentation with the help of wild yeast (starter).Whole rye flour and various seeds are stone grounded and sesame seeds are added. It has a lot of vitamins and fibe.

Finland

Karelian pies has a thin crust (typically made with rye flour) stuffed with potatoes, rice or carrots. A mix of butter and hard boiled eggs as topping is also relished.

France

Croissants are commonly eaten with various fillings.

Germany

Pretzels have a crispy crust with a chewy dough inside it. It is topped with ham and cheese or pumpkin seeds. Another famous dish is spaetzle, which is a type of pasta made from fresh eggs along with a variety of sausages which include bratwurst (made from minced pork and beef) and blutworst (made from blood sausage).

Currywurst sausage are available with or without the skin of the sausage and is served with ketchup and curry powder which are sprinkled on the top. Germany is the most populous country in the European Union with more than 80 million inhabitants and has a variety of regional cuisines. The traditional cuisine of the south-west includes plenty of white bread and noodles, whereas many traditional dishes of the Baltic Sea coast include potatoes and a variety of spices and seasonings. Along the Baltic and North sea coasts fish is eaten which is found in many traditional dishes. In Bavaria, meat and meat products, in particular pork is eaten (Hirschfelder and Schönberger, 2005). North-west of Germany has been influenced by the belgian cuisine, whereas the east shows polish influence and in the regions close to the czech border influences of the czech cuisine is observed (Hirschfelder and Schönberger, 2005). Sausages are very popular and a variety of sausages with different seasonings and flavours are available throughout the country. There are many varieties of bread, which form an important component of the German diet. The breads are typically based on rye and/or wheat and are solid and dark. Apple desserts, like apple cake, apple pancakes and apple strudel are popular. Stollen (sweetened yeast bread containing nuts and fruit) and lebkuchen (richly spiced ginger biscuits sweetened with honey) are eaten. Selected German traditional foods are black forest smoked ham. This ham has been produced for centuries in the black forest region and the recipe has been used through generations. The meat used is leg of pork which is cured with salt and herbs. This is then cold-dried and smoked using fir wood. Thuringian fried sausage is a long (20 cm), thin (2.6-2.8 cm diameter) fried sausage made of natural gut filled with pork meat. Swabian ravioli are a specialty with centuries of tradition. Maultaschen are quadratic or half-moon shaped and comprises of two-layer pasta dough forming a bag. They are usually filled with seasoned ground meat and spinach. Dresden fruit loaf is a dish in which a fruit loaf is baked. It is an old tradition in Dresden, a city situated in east Germany. The history of the Stollen can be traced back to the 14th century which is a sweet pastry, made of yeast dough, raisins, almonds, candied lemon and orange peel. Pumpernickel bread is the most famous and typical German bread and has been baked in the north-rhine westphalia region for many centuries. It has rye as the base with sourdough and is extremely dark and aromatic.

Greece

Gyros is a gyro flatbread which is stuffed with rotisserie-grilled meat. It is topped with creamy tzatziki sauce and relished by people. Traditional Greek cuisine are from the Hellenistic and the Byzantine periods. The cuisine has also incorporated from other civilisations, like the persian, the roman and the ottoman.Many names of Greek dishes resemble the cuisine of Turkey, Arab or Persia which include, moussaka, baklava and tzatziki. Even today the Greek cuisine has aspects of traditional practices in their cuisine. The traditional Greek diet is very healthy as they include a number of fruits, vegetables, legumes and cereals. Use of olive oil is common in the preparations. Fish and seafood are consumed frequently, meat, particularly of lamb, goat or pig is consumed. Wine is consumed regularly but in moderation. Dairy products are usually consumed in the form of cheese and yogurts; feta cheese is a world famous traditional greek food. Leek sausages are eaten since ancient times. The preparation uses leeks and has a distinct flavour due to the addition of many spices. Rabbit stew (*Kouneli stifado*) the word stifado comes from the ancient greek word tifos, meaning smoke or steam in which the meat is simmered with onions and various seasonings in tomato sauce. The cultivation and recipes of chickpeas in Greece goes back to the 3rd - 4th millennium BC. The chickpea soup is very common and includes ingredients like; chickpeas, water, onions, olive oil and lemon juice. For wine making fresh must of grapes is used.A variety of sweet dishes like marmalades, spoon sweets, petimezi (thick syrup of condensed must), must cookies and must jelly are eaten. This must jelly is made from 'cut' must, semolina, almonds, flavoured with cinnamon and topped with sesame seeds. Cherry tomato of Santorini has been common since the end of the 19th century. The local environment, the genetic profile of the specific cherry tomato plant and the empirical agricultural methods developed this distinctive food.

Hungary

Goulash is a meat stew with spices. The seasoning is made with paprika mixed with tomatoes, garlic, carrots, potatoes and beef. A savory version is "kettle goulash", which is made by frying beef or mutton with onions in lard.

Iceland : Being in the south of the arctic circle the people are Fighting for survival in many ways (Jónsson, 2005). As rainfall is high and the summer months are short the vegetation is subarctic. This has mainly grasses and lichens with limited woodland (1%). The cuisine therefore, mainly consists of meat (lamb, lamb offal), fish (stockfish, shark, skate) and dairy products (soft-cheese, whey). Until the 19th century the cultivation of vegetables (only cabbage, turnip, rutabaga and potato), cereals (any supplies had to be imported) was not much and the only fruits growing were wild berries. Traditionally icelandic moss and dulse became common food. The traditional preservation methods commonly used are drying, pickling in acid whey, fermentation and curing. The typical taste of traditional icelandic foods gets influenced by the preservation method used (Jónsson, 2005). Fish is eaten dried. A popular traditional meat product is smoked lamb (Hangikjöt). Dairy products also play an important role in the Icelandic diet.The only traditional cheese, Skyr (skimmed milk curd), is similar to thick yoghurt. It is usually eaten during breakfast or as a snack.Cured Greenland shark and smoked lamb are the traditional food. Among the vegetables, boiled potatoes in white sauce with red cabbage and canned green peas are traditional. Pickled blood sausage in which the whey is used for pickling and for food preservation. Sheep products are also eaten with oat meal porridge or with skyr, which is a yoghurt like product. On the fried product sugar is sprinkled on the top and served with potatoes and mashed swede. Skyr is a type of fresh cheese that evolved in Iceland to preserve milk and enhance the shelf life. It is made from skimmed milk, leaving the cream to make butter. Skyr is a popular traditional food for centuries it has been one of the most commonly consumed dairy products in Iceland.

Italy/Italia

Italian food has the best pizza and pasta. It is a flat bread garnished with tomatoes, mozzarella and basil. Another famous dish is latvia made out of dried peas, cooked with onions and fat.Italian cuisine has experienced influences from neighbouring regions. The Italian cuisine goes back to the 4th century BC. Stockfish, is very popular in Italy. Commonly used preservation methods in the middle ages were curing, drying and the use of brine and salt. The northern region has the influence of Romania and Germany and the southern parts by the Arab, now known as Mediterranean cuisine. Tomato, is the most important ingredients in Italian cuisine. This was introduced into Europe by the spanish from the americas. As it grows easily in mediterranean climates it has become very popular and is incorporated into almost all the local cuisine. North Germany does not use much of olive oil, pasta or tomato sauce as compared to the southern regions of Italy. Butter, rice, corn, cheese and cheese sauces are the preferred foods in this region. Central Italy is known for its olive oils, cured meats, cheeses (mainly from sheep's milk) and rich tomato sauces. Beef dishes are consumed more often than in other regions of Italy, and wild boar is very popular in the hills of Tuscany and Umbria. The south of Italy offers countless types of pasta, rich and spicy tomato sauces and pizza. In the north, fresh water fish such as perch, white and salmon trout are consumed (Demetri and Nascimbeni, 2008). The use of fresh, dried, ground or grated herbs and spices is common. Basil is the main herb in pesto sauces and oregano is used. Use of tomato sauces on pizza and rosemary with meat are common. A mixture of onion, garlic, carrot, celery and oil (usually olive oil), called soffritto, is the first step in the preparation of many dishes, such as stews, soups and sauces. The best traditional foods are Ricotta stuffed roll/ cannoli siciliani which consists of tubeshaped shells of fried pasta, filled with sifted sheep ricotta combined with vanilla, with chopped extra dark chocolate and pistachios This is flavoured with marsala wine or other flavourings. In the Catania area chopped pistachios are used for final decoration, in palermo the cannoli are decorated with fillets of candied orange peel. Vicentina cod/Bacala' is air-dried cod without addition of salt. Dry fish has been a good alternative to fresh fish as it is not perishable. Traditionally, Bacala' alla Vicentina is a dried cod stuffed with onions, flour, grated cheese, finely chopped parsley and sardines.This is cooked in milk and olive oil is served with slices of polenta gialla (yellow corn porridge). Pizza Napoletana Margharita was created in 1889 as a tribute to the Queen of Italy, Margherita di Savoia, on a visit to Naples. The authentic recipe for Pizza Napoletana Margherita includes local ingredients like san marzano tomatoes and mozzarella campana, made with milk from cows or buffalos raised in the plains of campania. The base is made of wheat flour, yeast, water and salt. Braised beef with Barolo wine is a typical dish of the Piedmont culinary tradition in which before being cooked, the meat is soaked in about 2 litres of Barolo wine for 12 h along with the vegetables and herbs. Later it is cooked slowly.

Lithuania

Cepelinai is a large oval dumpling which is made with potatoes and filled with ground meat, mushrooms and sometimes curd. Tuscan costagnaccio is a traditional dessert which uses the chestnuts and chestnut flour. Together with water and olive oil a basic dough is made and optional ingredients like pine nuts, raisins, walnuts, orange peel or rosemary are added. Mushrooms, berries, wild fruit and nuts are obtained from the forests and are used in many recipes (Imbrasiene, 2005). In the north-west region porridge and gruel are commonly consumed, however, cuisine of the central and north-east regions include a variety of pancakes and dishes made of cottage cheese, whereas in the south-east countless dishes based on buckwheat are prepared. In the south-west of Lithuania, smoked meat dishes and fatty pork are foods commonly eaten. Along the sea coast, fish is commonly consumed (Imbrasiene, 2005). Other traditional foods are the Cheese Džiugas/Sūris which is a medium-fat (~40%), very hard cheese, made from homogenised, pasteurised cow's milk, treated with enzymes and then ripened. It is produced in the western part of Lithuania (Žemaitija), especially in Telsiai. Cold fresh beetroot soup/ Šaltibarščiai recipe was documented in 1936. Cold beetroot soup is widely eaten in all lithuanian counties. It is made of milk, kefir, beetroot, sour cream, eggs, spring onions and fresh cucumber and served with potatoes. Boiled 'banger'/Kaimiškos dešrelės is a product widely eaten throughout Lithuania, especially in the Suvalkija region. They are made of beef and pork meat and lard.This is seasoned and filled into pig or sheep gut. They are then boiled for 20-25 minutes and eaten immediately after cooking. Zeppelins with meat/ Cepelinai has been a Lithuanian national dish. It is made from grated potatoes, usually containing ground meat. Sometimes dry cottage cheese (curd) is used instead. The potato dish resembles a Zeppelin in its shape, and is served with sour cream sauce or bits of bacon. Lithuanian biscuits 'Twigs'/Žagarėliai are widely eaten throughout Lithuania, especially in the Aukstaitija and Dzukija districts. They are delicate pastry biscuits that have been deep fried in fat. Lard or oil are usually used for frying. The dough is made of flour, butter, eggs and sugar.

Malta

Traditional Maltese rabbit stew, or Stuffat tal-Fenek, has been famous since the 11th century.

Netherlands

The stroopwafel is filled with syrup that comes in all kinds of sticky-sweet flavors from caramel and honey to wild berry and chocolate.

Poland/Polska

Pierogies from Poland are semicircular noodle dough dumplings often filled with fresh quark, boiled and minced potatoes and fried onions.Stuffing of sauerkraut, ground meat, cheese and fruits is also common.

Traditional Polish cuisine is influenced with the traditions of the neighbouring nations like Lithuania, Czech Republic and the Ukraine (Krzysztofek, 2005, Lemnis and Vitry, 1981; Adamkowska, 2008). Potatoes are an important staple diet since the second half of the 18th century (Adamkowska, 2008). Polish cuisine was formed at the beginning of the 19th century (Adamkowska, 2008). According to Krzysztofek (2005) many traditional foods have been preserved. One of the national dishes is Bigos (type of stew) and the typical ingredients include sauerkraut, different meats, sausage, dried mushrooms and prunes. Pork is eaten as fried pork with boiled potatoes and white cabbage. Consumption of bread, Kashas (grits) and other cereal-based dishes, dumplings and potatoes are very common. Pickled foods such as vegetables (cucumbers), fish (herrings) and wild mushrooms are popular. Traditional cakes include gingerbread, poppy seed cake, faworki (crisp cakes), easter mazurkas, doughnuts and sękacz (tree cake).Chlodnik/Chlodnik cold soup is very popular in the Polish cuisine. This soup is either made with soured beetroot juice or the juice of soured cucumbers, which is mixed with sour cream or sour whole milk. Minced dill and chive are used for seasoning. Pork chop is the most popular meat and traditionally prepared by dipping in flour with eggs diluted in water and bread crumbs. They are then fried and served with cabbage and potatoes and topped with fat. Stew made of sauerkraut, meat and dried mushrooms is known as Bigos and is one of the most popular dish. Onions, dried mushrooms, red wine and different herbs and spices are also added to it. Prunes give this dish a slightly sweet taste. Tree cake, locally known as sękacz has been present in Polish recipe collections from 19th century.Smoked ewe's milk cheese known as oscypek is a typical cheese made up of ewe's milk. It has been a valuable source of energy. The cheese is put into brine and then smoked for the typical taste and colour.

Portugal

These are custard-filled sweet pastries which were originally prepared by the monks at the Jerónimos monastery and since then it has become very famous in portugal.

The cuisines have been influenced by other costal cuisines (Costa, 2005). Many of the typical Portuguese pastry varieties originate from monasteries and convents, like Pastéis de Belém and Queijadas de Sintra, which dates back to the 13th century. Also, many desserts, cakes and pastries rich in sugar illustrate the arabian influence on portuguese cooking.The use of almonds, honey and orange blossom has moorish influence (Costa, 2005).Olive oil is used for cooking. It is also used for salad dressing, especially for boiled legumes and fresh salads. Another food that is consumed is salt cod, which is a staple ingredient in many dishes (Costa, 2005). The cuisine across regions and islands is varied. Residents of Porto are known as Tripeiros (tripe eaters) since centuries. In the central and southern part, people eat local fish and vegetables. The traditional dishes include a wide variety of tropical fruits and local meat and fish.The use of traditional cheeses made from sheep or goat's milk, like Queijo da Serra are used. Among the selected traditional foods; Green kale soup/Caldo verde, which is a potato-thickened soup made with chopped kale and seasoned with a dash of olive oil and a slice of Chouriço. Due to its simplicity and lightness, it is usually served at the beginning of a meal or as a late supper. The main ingredients of this soup are potatoes, galega kale, onion, garlic and chouriço, and it is often served with corn bread and is known as the national dish. Cod with chickpeas/Bacalhau com grão is a dried salt cod which is traditionally served with chickpeas and potatoes, sprinkled with onions, garlic and parsley. It is seasoned with olive oil, vinegar, pepper and paprika (optional). It is garnished with boiled eggs. Portuguese boiled dinner is a robust one-dish meal of boiled beef, chicken, pork and smoked sausages along with a variety of vegetables. Oven-roasted goat kid/Cabrito assado no forno, is cooked in a wood fire oven.The goat kid is placed on a bay stick grill which contains the rice and the broth. The unique taste is obtained by the meat juice falling on the rice during cooking. Egg sweet from Murça is a dish that contains rich egg-yolk and almond sweets. It is prepared using iron pan over a wood fire to cook.

Romania

Cabbage Rolls are made out of minced meat with rice and spices that are rolled into fermented cabbage leaves. Then these are boiled for hours in a special sauce that provides flavors and tenderizes the meat.

Spain/España : Potatoes, tomatoes, courgettes and peppers are widespread throughout the country and are the main ingredients of a large number of traditional dishes. The spanish traditional cuisine has ingredients like olive oil, tomatoes and a wide variety of fruits, vegetables and legumes. The types of meat consumed, varies depending on local farming and regional traditions. However, the commonly consumed meats are chicken, pork, lamb and veal. Fish and seafood are also regularly eaten in many parts of Spain. Among the flavourings, garlic, is used in numerous traditional dishes. Meat, fish pies, scallops and veal are popular. Eastern part is known for its legume dishes and a strong blue cheese (Queso cabrales). In the Basque country, fish dishes play an important role in traditional eating habits. Rice is a staple food and the world famous variety paella comes from this region. The southern part is an arid region, with many olive trees and grape vines (Pappas, 2008b). Tapas, little snacks in between or before a meal, typical of Spain are popular. Traditional Spanish dishes include Tortillas (omelettes, especially potato omelettes), pulpo a feira (Galician style octopus), pescaíto frito (small fried fish) or pimientos de padrón (miniature roasted green peppers served with olive oil and salt). Hot vegetable sauce/Mojo picón Mojo picón comes from the Canary Islands, where it is a staple food and is eaten with papas arrugadas (potatoes boiled in salt water). The sauce is spicy made up with a paste of garlic, cumin, sweet paprika, chilli pepper, sea salt, olive oil and vinegar. Cardoon in almond sauce is a traditional dish mainly containing cardoon and pieces of pork belly in a sauce of almonds, seasoned with garlic. 'Escalivada'/Escalivada is a roasted pepper and aubergine salad which is a typical catalonian dish. It has many different types of grilled vegetables. It is usually made of aubergines, sweet red peppers, tomatoes and sweet onions. Only olive oil and salt are used for seasoning. Galician octopus is a typical dish with main ingredients of octopus, potatoes and onions.It is seasoned with salt and sweet and spicy paprika. Almond cake is a typical dessert from Las alpujarras in Granada.These are flavoured with aniseed, cinnamon, sesame, almonds and honey. The basic ingredients of the cake are almonds, sugar and egg white.

Slovenia

Bryndzové Halušky is a popular national dish of Slovakia. The potato-dough dumplings (halušky) are filled with sheep cheese (bryndza) and sprinkled with bacon or sausage on the top.Another famous dish is a crepe-like dish called Štruklji. These are cottage cheese pancakes made with tarragon.

Spain

Paella is a rice-based dish, which comes with either vegetables, meat or seafood.

Sweden

Cinnamon rolls are baked with large amounts of cinnamon with a topping of crunchy sugar.

Turkey: The Turkish cuisine got initiated during the reign of Fatih Sultan Mehmet (Mehment II the Conqueror) (Baysal et al. 2006). The cuisine is influenced by the arabic cuisine (Baysal et al. 2006). The Turkish cuisine has some common specialities that can be found throughout the country like butter, yoghurt, cheese, honey, meat and cereals are local foods. Kebab, Börek, meat, vegetable dishes and Helva desserts are eaten. Olive oil is used both in hot and cold dishes.A small fish, the Hamsi, is a traditional dish made with Hamsi, tomatoes, potatoes, onions, and lemon, cooked with olive oil and served as a main dish for lunch or dinner (Sancar, 2005).Turkish traditional dishes like Pilaf, use currants, cinnamon, pine nuts, chilli peppers, mint, parsley, dill / cumin as flavourings of meats and seafood. Among soups tarhana, rice, lentil and offal are popular. Commonly used vegetables are aubergines, artichokes, beans, beetroot, chard, chick peas, cucumbers, mushrooms, onions, peppers, spinach and tomatoes. They consume these vegetables as Dolmas (stuffed vegetables) which is consumed with yogurt. Pastirma is pressed meat. Excess meat is stored in brine or by drying techniques. It is prepared using veal, flour, hot chilli powder, dried garlic and salt. Baklava Baklava is one of the most famous Turkish desserts and is consumed throughout the country. The dough of this sweet pastry is made of wheat flour and eggs and the filling is a mixture of sugar, semolina, milk and pistachios. Before baking, melted butter is poured over the Baklava.

Beverages around the globe

Belgium is famous for beer.The country has over 150 different breweries. Alcoholic beverages contain ethyl alcohol in varying proportions. Beer contains 2-5% and wine 8-10% of alcohol, while brandy, rum and whisky contain much higher concentrations (30-40%). Alcohol has been extensively abused as an appetite stimulant and as a sedative-hypnotic drug. Alcohol intake, which is initiated as an innocent social habit may gradually result in a serious addiction. It may lead to several serious psycho-social problems and accidents. Alcohol provides higher calories (7 Kcal/g) than carbohydrates and proteins and can therefore, contribute to obesity. Ironically, excessive intake of alcohol is known to suppress appetite and interfere with absorption and metabolism of nutrients, leading to various nutritional deficiency diseases. Excessive intake of alcohol suppresses appetite and as a result, leads to several nutritional deficiency diseases. People who regularly consume more than 60 ml of alcoholic drinks are at a higher risk for hypertension and stroke. Alcohol intake has also been shown to increase the risk of cancer of the mouth, larynx and oesophagus prostate and of the breast in women. Excessive alcohol intake weakens the heart muscle (cardiomyopathy) and also damages the liver (cirrhosis), brain and peripheral nerves. It also increases serum triglycerides. All these food items have been the traditions or customs / beliefs which run in the generations in each respective place and plays an important role in any cultural identification. It can be observed that each region has a different preparation method and different recipe of food items. It has also been observed that the basic cereal and other ingredients used are mainly depending on the staple food of that particular area and have been consumed locally and regionally for an extended time period. Preparation methods of traditional foods are part of the people of a region. It is therefore that each culture, ethnic group or region has specific traditions. Sometimes, the traditions like religious customs, overlap different cultures, ethnic groups or regions. The food ingredients in food preparation and methods are very specific in different areas which are passed from one generation to the next. Therefore, these are termed as 'traditional foods'. These foods have played a major role in traditions of different cultures and regions for many years. Importance should be given to them or they might disappear in time to come. It is therefore, vital to study and document traditional foods to sustain important elements of various cultures. There are enormous number of traditional foods around the world. It is due to the globalisation of traditional foods that the consumers' perception of traditional foods has been changed. This has helped the people of different regions to adopt foods from other regions and consume a mixed regional meals for nutritional security. According to EuroFIR, European Food Information ReSource (EuroFIR) Network of Excellence the definition of traditional foods has been towards the establishment of a pan-European food information reSource in the form of a portal, which is helping people to access the food composition data across Europe online and around the world. The importance of including composition data for traditional European foods has been recognised, and a work package concentrating on traditional foods has been set up. A first step of the EuroFIR Traditional Foods work package has been the development of a definition of the term traditional foods (Trichopoulou et al. 2007). As a clear definition it is important to select and further collect information about traditional foods. This includes ingredients, preparation methods and food composition data.The EuroFIR definition of traditional foods has been acknowledged by the Food and Agriculture Organization (FAO, 2008) at the FAO Regional conference for Europe (26th) in Innsbruck, Austria, 2008. Accordingly, the definition was given as the food conforming to establish a practice or specifications which was present prior to the second world war. Traditional food is a food with a specific feature which distinguishes it clearly from other similar products of the same category in terms of the use of traditional ingredients (raw materials of primary products) or traditional composition or traditional type of production and/or processing method as given in the definition.Traditional ingredient (raw material or primary product) are the raw material (species and/or varieties) or primary product, alone or as an ingredient which have been used in identifiable geographical areas and should be in use today. Its characteristics must be in accordance with the current specifications of national and EU legislation.The uniquely identifiable composition/ingredients when necessary are differentiated from the composition defined by the generally recognised characteristics of the wider food group to which the product belongs. In traditional type of production and/or processing it should remain in line with methods which were used originally and that the food's intrinsic features such as its physical, chemical, microbiological or organoleptic features should be maintained.The standard procedures for a systematic investigation of traditional foods across Europe were developed and then applied in the participating countries. The selection process was based on the three main steps: documentation, prioritisation, evaluation and selection (Trichopoulou et al. 2007). Each selected food was prepared using a traditional recipe, traditional ingredients and traditional preparation methods from the region it is traditionally consumed. Traditional techniques and recipes were followed in all Europe situations. The recipe and preparation methods were documented on paper and with pictures though the video-taping was optional. With all the above, now a systematic study of traditional foods basically for documentation, prioritisation, evaluation and selection have been made mainly to represent the starter, main dishes and the dessert. In a data study of traditional foods of Europe the following have been selected.

Traditional foods in European food composition databases are used for many purposes, such as food labelling, nutrition and health research and policy making. Despite increasing globalisation, traditional foods still contribute a fair amount to food intake in most European countries. It is essential to have information about the macro and micronutrient composition of traditional foods. The information will be useful for the determination of the role of traditional foods in the dietary patterns and nutrient intake by a population. However, many national databases are currently lacking in nutrient data on country-specific traditional foods (Trichopoulou et al. 2007).To fill these gaps, traditional foods need to be systematically investigated and information on their nutritional composition must be included in national food composition tables and databases. Further information on the use of food composition databases can be found in the 2nd Synthesis Report namely, The different uses of food composition databases (Williamson, 2006).

Looking into the health aspects of traditional foods it is found that there is a diversity of national and regional traditional foods which are made with a variety of ingredients. This usually reflects the traditionally produced local ingredients. Due to huge variations in the cuisines across European countries and regions, it would be misleading to make general statements about the association between traditional foods and health that is valid for all European traditional cuisines. The fact that foods are traditional does not automatically mean that they provide any particular health benefits. The impact of traditional foods on our health depends on their nutritional composition. A cuisine including high amounts of starchy foods, fruits and vegetables and moderate amounts of fish and meat will provide more health benefits than a dietary pattern which is high in meat and fat and low in fruit and vegetables. Traditional Mediterranean diets incorporating a high proportion of fruit and vegetables, olive oil and a relatively high consumption of fish are considered healthy (Trichopoulou et al. 2006). In these countries, traditional foods are thus associated with better health. However, in other countries traditional foods may be perceived as unhealthy. In European countries where the traditional cuisines are very high in meat and fat, then, they will be considered unfavourable from a health perspective. Therefore, to draw conclusions about the health benefits of a certain cuisine or the traditional foods it becomes difficult. Some traditional cuisines, with a high proportion of animal products and fat, could have advantages in the past.In northern part of the world due to low temperatures, energy expenditure during winter times is higher. In such situations the traditional foods have been crucial for the survival. The profession like farming or as labourers which involves more physical activity needs more energy per day. In such situation intake of the fat content in the diet is helpful.The nutritional requirements change on the type of working schedule of a person. Recently, it has been found that the average energy requirements are significantly lower than in the past centuries. To maximise the health benefits of a diet, it should therefore be adapted to the nutritional needs in a population. Some of the traditional foods which were developed a long time ago have their place in a healthy diet today. However, many of them do not meet the nutritional needs as they are too high in energy or fat.

Recently, it has been observed that many of the traditional foods are being modified based on the nutrient composition to make them more appropriate for the 21st century. However, that could mean that these foods are not considered traditional anymore.The Mediterranean diet, has been a very beneficial diet to health. This has been a model with favourable compositions of some traditional cuisines, along with utilization of the local ingredients.Due to the increase in global exchange for many products including foodstuffs, the culinary borders are not very defined. Food habits in many European regions have changed considerably since the second world war which is influencing the traditional cuisine of a country or region. Particularly, within the modern day Europe, the food market is trying to open up avenues for the development and expansion of the European Union. Internationalisation of nutrition habits can be observed, due to the worldwide trading and the assimilation of new lifestyle habits which are being adopted from other countries around the globe (Besch, 2002). Studies carried out in four European countries (Germany, Great Britain, France and Spain) examined the internationalisation of nutrition habits (Ziemann 1999; Besch, 2002). Alderman et al., (2006) have discussed on the long term consequences of early childhood malnutrition. It was found that the traditional food products and meals are becoming less important. It is the processed foods which are replacing the traditional foods. Eating as a family is becoming less frequent. Meals are being replaced by snacks. Out-of-home consumption has been on the rise in recent times. An increased internationalisation has been found in the urban areas. The biggest changes and internationalisation has been found in Great Britain. In Germany and France, the change was of a lesser degree with hardly any changes in Spain (Besch, 2002).According to Trichopoulou et al. (2007), dietary patterns are influenced by the local availability of foods and the cultural and socioeconomic environment, but there is a trend for transfer and assimilation of new habits between countries. In the 1960's the diet of mediterranean populations was characterised by a high consumption of fruits and vegetables, as opposed to the low consumption of these foods in northern European countries. These large differences seem to be diminishing and contemporary patterns reveal that the mediterranean populations are straying from their traditional diet, whereas in north European countries mediterranean-style eating has increased in popularity (Trichopoulou et al.2007).Because of the increasing globalisation and internationalisation of the food market, many traditional foods might have a risk of disappearing. Therefore, it is important to document the traditional foods and recipes which are an important part of cultural heritage.Traditional foods within the EuroFIR (European Food Information Resource) network involves the funding given under the EU 6th framework for a programme on food quality and safety. The work was taken up in 2005 till 2009. From 2010, EuroFIR is called as EuroFIR AISBL and it is self funded (www.eurofir.net). The aim has been to provide food information to different stakeholders in Europe. One of the main objectives has been to create a pan-European food information reSource in the form of a portal, allowing access to online food composition data across Europe, by linking national food composition databases (FCDB) throughout Europe. Links have also been to establish them with countries outside Europe. Traditional foods play an important role in dietary habits of Europeans. Therefore, food composition data from traditional foods are necessary. To fill existing gaps, one of the work packages within the EuroFIR project has been the analysis, documentation and selection of the traditional foods from the countries participating in this work package. Such documentation and analysis of traditional foods is going to be very valuable for the food industry. The obtained information and data can be used for the development of new food products based on traditional recipes. EuroFIR has been particularly working with small and medium sized enterprises (SMEs). The EuroFIR work package known as 'Traditional Foods' includes thirteen european countries. The program is led by Dr Helena Costa from the National Institute of Health, Portugal (INSA).The program is also helping in establishing a common methodology for the systematic investigation of traditional foods across Europe. New data is being prepared on the nutritional composition of traditional foods for inclusion in national food composition tables with representative raw ingredients and recipes. It is a nework program and the partners in this work package include: University of Vienna (UVI)/Graz University of Technology (GUT), Austria Ghent University (RUG), Belgium National Centre of Public Health Protection (NCPHP), Bulgaria National Food Institute, Technical University of Denmark (DTU), Denmark The Federal Research Centre for Nutrition and Food (BFEL), Germany , Department of Hygiene and Epidemiology, Medical School, University of Athens (NKUA), Greece Matis ohs (MATIS), Iceland National Institute for Food and Nutrition Research (INRAN)/Centro per lo Studio e la Prevenzione Oncologia (CSPO), Italy , National Nutrition Centre

(NNC), Lithuania. National Food and Nutrition Institute (NFNI), Poland National Institute of Health (INSA), Portugal Centre for Superior Studies on Nutrition and Dietetics (CESNID)/ University of Granada (UGR) and Spain Tubitak Marmara Research Centre (TUBITAK), Turkey.

The EU has created product designations which are linked to geographical origin or traditional production methods. Sometimes, foods and food products of a certain region acquire a reputation extending beyond their production region, and they can find themselves in competition with products which pass themselves off as the genuine article and take the same name. This may discourage producers and also mislead consumers. In 1992, the EU created systems to promote and protect valuable food names.The systems have been updated and improved in 2006 (European Commission, 2007a). The product designations created by the EU fall into two categories: those linked to a geographical region or territory, and those relating to a particular production method. The designations linked to a territory are: Protected Designation of Origin (PDO) PDO means the name of a region, a specific place, or in exceptional cases, a country, used to describe an agricultural product or foodstuff: originating in that region, specific place or country and possessing quality or characteristics which are essentially or exclusively due to a particular geographical environment with its inherent natural and human factors.The production, processing and preparation takes place in the defined geographical area. PDO products require different stages of the food production process to be carried out in the concerned area. These could be like, Shetland lamb from the UK, Jamón Huelva (ham) from Spain, Allgäuer Emmentaler (cheese) from Germany or Wachauer Marille (apricot) from Austria. Protected Geographical Indication (PGI) PGI means the name of a region, a specific place or, a country which is used to describe an agricultural product or a foodstuff that is originating in that region, specific place or country and which possesses a specific quality or reputation or other characteristic attributes to that geographical origin.Also the production, processing and preparation of the particular product taking place in the defined geographical area. PGI products thus require that at least one stage in the production process must be carried out in that area, while the raw materials used in production may come from another region. Examples of PGI include Jambon d'Ardenne (ham) from Belgium, Esrom (cheese) from Denmark, Pomodoro di Pachino (tomatoes) from Italy or Džiugas (cheese) from Lithuania. A third designation is used for products that are linked to a particular method of production known as : Traditional Speciality Guaranteed (TSG). The specific features required by the TSG system includes the use of traditional raw materials in production of a foodstuff or traditional composition or mode of production and processing which reflects the traditional methods. Thus, to obtain the TSG designation a product must possess features that distinguishes it from other products and it must be traditional meaning proven usage in the EU market for a time period through generations. For the EU this means a minimum of 25 years. This is a significantly shorter period than suggested by EuroFIR. Contrary to the EuroFIR definition of traditional foods, TSG products are not linked to any geographical region and could be used by any producer. Some of the examples for products registered as TSG are Panellets (small biscuits or cakes) from Spain, Moules de Bouchot (mussles) from France, Pizza Napoletana from Italy and Kabanosy (meat product) from Poland. According to the EU, traditional food products are often the result of agricultural practices that preserve and enhance the rural environments. Their production is very much in line with the current EU thinking on rural development, preservation of biodiversity and sustainability (European Commission, 2007b).

A European project known as Truefood has the aim to introduce innovation into the traditional food production system and industry. It has an aim to bring research and the industry come closer for effective collaboration and technology transfer. As the focus has been more on production and industry, the definition given is on the traditional food products rather than traditional foods in general. The production must be local (national/regional/local). Once the production gets started in other countries, then the food is not traditional. It should be an authentic recipe (mix of ingredients) and/or authentic origin of raw material and/or authentic production process. Commercially, the product must be available to the public for at least 50 years (from 1950 or before) in stores or restaurants; it may happen that during that period the food product disappeared from the market, but was on sale for at least 50 years with gastronomic heritage. http://www.truefood.eu/

Another definition of traditional food products obtained through the qualitative consumer research was conducted in the course of Truefood during 2007 in which six European countries (Belgium, Italy, France, Spain, Poland, and Norway) participated and according to that "A traditional food product is a product frequently consumed or associated to specific celebrations and/or seasons, normally transmitted from one generation to another, made with care in a specific way according to the gastronomic heritage, with little or no processing/manipulation, that is distinguished and known because of its sensory properties and associated to a certain local area, region or country" (Vanhonacker et al. 2008, 2013 and Verbeke, 2005).

It has been observed that in India people often follow dietary restrictions based on their religion or faith. Among the hindu communities eating of beef is not allowed. Slaughtering of cows has been banned in many states of India (Cowed down by the ban. Hindustan Times, 2012). The community called as

Vaishnavas do not consume garlic and onions as given in the Bhagavad Gita (Bhagavad-Gita, 2013; Chapter 17, Verse 8, 9, 10). Jains follow a strict form of lacto-vegetarianism, known as Jain vegetarianism, which in addition to being completely lacto-vegetarian, also excludes all root vegetables such as carrots and potatoes because when the root is pulled up, organisms that live around the root also dies (Shah, 1998). Muslims do not eat pork or pork products. Unlike some East asian countries, dogs and cats are not considered suitable for consumption.There has been an unique palette in every culture when food is considered. The climate of the geographical region is very important. India, for the case being has a tropical monsoon climate which is conducive for a truly wide variety of vegetation to grow. India (or rather the Indian subcontinent) is the native of thousands of plant species, especially condiments and spices which helped in the preparation of many tasty cuisines.

8

Fortification in Traditional Foods and Nutritional Security

Food fortification is the procedure in which micronutrients (essential trace elements and vitamins) are added to the food. As a public health policy fortification can be carried out by the food manufacturers or by the governments to help people in combatting dietary deficiencies within a population. Fortification can be done in food even if these nutrients are originally present or not. However, it improves the nutritional status of a population by overcoming the deficiencies of micronutrients. Deficiency of micronutrients is known as hidden hunger which is a serious health risk. Access to safe and nutritious food is a must.Due to the lack of consumption of a balanced diet, adequate micronutrients are not supplied to the body. Therefore, there is a considerable loss of nutrients during the processing of food. One of the strategies to address this problem is fortification of food.

Nutritional Deficiencies: Significance of Micronutrients

The micro-nutrients can be vitamins and minerals which are required in very low quantities. They help the body to produce enzymes, hormones and other substances which are essential for proper growth and development. Inadequacies in intake of micronutrients cause deficiencies leading to malnutrition.(http://www.who.int/mediacentre/factsheets/malnutrition/en/).

More than 2 billion people in the world suffer from micro-nutrient deficiencies. These are mainly caused if the food intake of nutritious food is insufficient, intake of / an imbalanced diet and consumption of only one type of ingredient which may be lacking in certain micro-nutrients.Consumption of one staple diet as in the case of corn, continuous eating may cause illness like pellagra which is caused due to the dietary deficiency of Niacin or Vitamin B3. Corn is deficient in niacin and the amino acid tryptophan which is a precursor for Niacin. With the innovations in science new varieties of nutrition enriched maize are being produced. Many of the traditional foods have a food combination which balances many nutrients. Traditional Punjabi cuisine that consists of maize rotis and sabzi prepared from mustard greens (sarson da saag) has many vitamins and amino acids balancing the deficiency of niacin and tryptophan in the maize. Around the world there are many such traditional dishes. The biggest problem in India is of iron deficiency (anemia) which causes around 20% of maternal deaths directly and other 50% due to iron deficiency related problems. The output of individuals gets reduced. It is therefore, suggested, to consume vitamin A and iron rich foods. Another major health problem in India is the deficiency of vitamin D with a prevalence of 70% -100% among the general population. As it is very essential for the heart, brain and the immune functions it is important to consume it in required quantities. The deficiency causes bones to become thin, brittle, soft or unshapes them. As per the scientific data it has been shown that it plays a vital role in the metabolim of calcium and absorption and in maintaining serum calcium levels. The intestinal calcium absorption happens through the induction of calcium binding protein and by the stimulation of bone mineralization.

If the daily diet is incorporated with grains, pulses, nuts, tubers and fruits it helps in improving the micronutrient deficiencies. Recently, a lot of research has taken place towards food fortification, in which ingredients rich in micronutrients (iron, vitamins and minerals) are added. This is helping the population to a large extent. It has been well studied that if the mother is undernourished she gives birth to an underweight and may be a stunted child. Therefore, if malnutrition in mothers could be prevented then the disabilities in their infants could come down by one third. Iron, vitamin A, iodine and folate, play a vital role as micronutrients for the mother in child's development (https://www.unicef.org/role.html). The World Bank has been recommending the use of biofortified cereals, like Golden rice which is improving to benefit the human kind and the environment (Dubock, 2017). Agriculture has the biggest negative impact on biodiversity of cultivated land. However, intensive agriculture has maintained the global biodiversity by protection of wild lands from food production.Integration of conventional plant breeding, including mutagenesis, biotechnology and genetic modification, better soil husbandry and organic farming, agroecology management, pesticides, organic and inorganic fertilisers can definitely produce crops with better nutrients which will give healthy human beings.Per-capita food production has increased during the last four decades after the 1960's Green revolution.This has resulted in more rice production which was 34.6 million tons in 1960 which increased to 154 million tons in 2013–2014. Similarly the production of wheat also increased from 10.3 million tons in 1960 to 97 million tons in 2013–2014. During the same period, the population of India increased from 350 million to 1.3 billion. During 2013–2014, India exported 10.7 million tons of rice which was much more than any other country (Khush, 2015). More recently, after 1990 due to the knowledge of plant genetics and plant breeding many technology improvements have taken place which have helped towards higher yields (Dubock, 2013; Brookes and Barfoot, 2016). Often poverty is associated for limited dietary diversity. Carbohydrate-rich grains are an important energy source. Macronutrients like fats and proteins and micronutrients/minerals like iodine, iron, zinc and vitamins (vitamins A, C, D, B1 and B3) are required for healthy development and day today life activities. The micronutrients are present in the soil in the form of minerals and are taken up from there by the plants in which they get accumulated. Vitamins are synthesised either by the plants and also by the animals. Energy-rich grains are very important, but without dietary diversity they are not sufficient for a healthy life (Dubock, 2013).

Dietary diversity can come from plant products and also from the animal products. The people get benefitted both for macronutrients and micronutrients with such a varied diet which usually provides a balanced diet. It has been observed that it is the poor strata of people usually present in the developing countries who have very limited access to animal products due to economic reasons.Since the 1970's, fortification in food, first with respect to iron (Baker et al.1999), then folate (Crider et al.2011) and from the 1980's vitamin A (Semba, 2012) has brought global awareness to the public that dietary micronutrient deficiency can be prevented by fortification to improve human health. It will also bring down mortality rate. The United Nations and other supranational organisations have taken this matter and have addressed the problem of nutrition.It was in the year 2000, that heads of 129 states met at Johannesburg and ratified the Millennium Development Goals. They had an aim to achieve many of these goals between 1990 and 2015, in order to better the food nutrition for the poor around the world (United Nations, 2015). Despite the unity of intention, goal 1, emphasizes to eradicate extreme poverty and hunger. Goals 4 and 5, are to reduce the mortality rate among child and mother (United Nations, 2015). According to Juma, Africa and most developing countries need a translational application of already established effective technologies and not basic research. Recently, in Africa and many other countries the mobile telephone is being used to broadcast the applications been developed (Juma, 2016). In the 2016 G20 summit, the agriculture ministers has shown their concern that even after tremendous efforts, 795 million people in the world still suffer from chronic hunger and 2 billion people from micronutrient malnutrition (G20 Ministers, 2016).

Vitamin A deficiency

Human populations deficient in the minerals such as iron, zinc and in vitamin A, suffer from a variety of serious developmental and public health issues and these deficiencies are more widespread globally than other important micronutrient deficiencies (Dubock, 2013). Vitamin A deficiency, VAD, which mostly affects the children of less than 5 years old, and to a lesser extent their mothers, is widespread and well documented by the World Health Organisation (WHO, 2009).

By the 1990's, VAD was known as the major cause of childhood blindness globally, with about 500, 000 cases every year. Out of this, about two-thirds die in case they were not treated (WHO, Whitcher, 2001). As a result, for more than a quarter of a century, vitamin A deficiency (VAD) has been recognised by the United Nations as a significant public health problem. Key milestones included the:1990 UN World Summit for Children, where 50 heads of government and senior government officials committed their governments to the virtual elimination of VAD by the year 2000 (UNICEF, 1990). In the UN International Conference on Nutrition, (1992) it was concluded that VAD control is the most cost-effective child health/survival strategy. The governments must pursue and all sectors of the society should support the virtual elimination of VAD. They looked into the strategies which included the promotion of breastfeeding, dietary diversification, vitamin A supplementation and food fortification. They also recommended that the locally available food-based strategies should be the first priority. However, it was felt that Vitamin A capsule supplementation can be only an interim measure (FAO of UN, 1992). It was in the year 2004, that UNICEF along with the Micronutrient Initiative made a report on "Vitamin and Mineral Deficiency", which concluded that controlling vitamin and mineral deficiencies can improve the lives of people (UNICEF, 2004). According to Semba, (2012) VAD suppresses the human immune system and causes a nutritionally acquired immune deficiency syndrome. Eyesight problems and immune deficiency-related deaths are two different morbidities of vitamin A deficiency. Most who die as a result of VAD do not become blind first. The scientists (West et al., 2010; Dubock, 2013) have found that the increased susceptibility to disease as a result of VAD results in the majority of the millions of preventable <5-year-old child deaths, mainly among children, annually. A universally available source of vitamin A could save 23–34% of all deaths of children under 5-year-old globally and reduce measles mortality by 50% (West et al., 2010, Mayo-Wilson et al, 2011; WHO, 2013). UN regularly measures and publishes global all-cause child mortality (UN, 2010). The mortality due to VAD as compared to other public health mortality is found to be more in the countries where poverty is widespread.

Table 10: Global mortality from important public health diseases

Global mortality (millions)	2010	2014
Vitamin A deficiency	1.9–2.8	1.4–2.1[a]
HIV/AIDS	1.8	1.2[b]
Tuberculosis	1.4	1.1[c]
Malaria	0.7	0.6

Data Sources: United Nations

Among children < 5 years; one in three HIV deaths is due to TB. The meta-analyses of the effects of a Source of vitamin A which have led to these estimates are based on research with vitamin A supplementation which cannot be repeated (Table 10). Because the already proven benefits of vitamin A supplementation are clear it is not acceptable to have a control group (vitamin A deficient and without a source of vitamin A) in new research. There are no new data but only new analyses.Although MDG Goal 1 was not achieved, it is clear from UN reports that there has been a continuous reduction in child mortality since the 1990's (Dubock, 2017) and it is clear that improved living standards, for example the availability of clean water, as well as vitamin A capsule supplementation programmes, have saved the sight and the lives of millions from the effects of vitamin A deficiency. Yet vitamin A deficiency remains a very significant public health problem (West et al.2010; Stevens et al.2015).It is principally associated with poverty and is preventable at low cost (FAO of UN, 1992). Semba has noted that for children who are < 5-year-old, the child mortality in India, may be due to the low coverage of India's vitamin A supplementation programme which is equivalent to the total child mortality of the sub-Saharan African countries (Semba, 2012; http://www.who.int/maternal_child_adolescent/documents/levels_trends_child_mortality_2015/en/). VAD is a challenge as it is only animals including human beings who are able to synthesise vitamin A. It is not synthesized by the plants. The direct sources of vitamin A are animal products which include; liver, butter, milk and eggs. Animals and humans synthesise vitamin A from carotenoid chemicals naturally occurring in coloured fruits, vegetables and leaves. Beta-carotene (β-carotene) is an important provitamin A carotenoid which is found in all coloured plant foods. In developed countries, VAD is not a big problem because the diet usually includes animal products and various sources of provitamin A carotenoids. Fortification of food products is also very common. Taking vitamin supplements by newborns, pregnant and lactating women is also common. However, in the poorer sections of the developing countries, food sources are most valuable in terms of micronutrients. For food security, staple crops like rice are cheaper and therefore makes up most of the diet.

Energy source and food budget of various staple foods is given in. It shares the energy source and food budget in rural Bangladesh. Source: www.harvestplus.org/content/food-crisis. The bioavailability of vitamin A from fruit and vegetables is very limited which increases the problem of VAD (Tang, 2013). It has been estimated that young children between age 1 and 3 years would need to eat eight servings of dark green leafy vegetables per day in order to meet the recommended dietary allowance (RDA) for vitamin A. This observation has brought to the conclusion that it is extremely difficult for most poor and young children to meet their vitamin A requirements only through vegetable and fruit intake (Semba, 2012). According to Bouis et al. (2012) there are three and a half billion people daily who eat rice and therefore, they obtain most of their calories from rice. This accounts for 80% of their calorie intake in Bangladesh. Dubock (2014a,b) found that the countries where the diet lacks in sufficient vitamin A have high rates of infant and maternal mortality.

There are several strategies for addressing micronutrient deficiencies caused by insufficient dietary diversity. Industrial fortification (iodine to salt, vitamins A and D to margarine, fluoride to toothpaste, folic acid to flour) have been successfully implemented to ensure micronutrients sufficiency to the populations. Providing tablets, syrup or capsules containing micronutrients and their consumption has been employed in both industrialised and developing countries to combat the deficiencies. However, for fortification and supplementation some level of manufacturing, distribution and infrastructure are required. Therefore, it becomes difficult to reach the needy. Distribution of vitamin A capsules was introduced to vulnerable populations in the early 1990's in order to combat vitamin A deficiency (Semba, 2012 and West et al.2010). According to WHO, vulnerable children between 6 months and 5 years having VAD should be given a capsule every 6 months.

(WHO.Nutrition, micronutrient deficiencies, vitamin deficiency.2013. http:// www. who. int/ nutrition/ topics/ vad/en/., WHO. Guideline: vitamin A supplementation in infants and children 6–59 months of age.2011.p1–25. http:// apps.who.int/iris/bitstream/10665/44664/1/9789241501767_eng.pdf., WHO. Guideline:neonatal vitamin A supplementation.2011.p.1–25. http://apps.who. int/iris/bitstream/10665/44626/1/9789241501798_eng.pdf). In studies done by Ye et al., (2000) and Potrykus(2014) proved that they could modify the genome of white rice to produce beta-carotene, the precursor which the human body uses as a source to make vitamin A. IRRI has studied the breeding of the selected nutritional trait, transformation event GR2E which is responsible for synthesising beta-carotene in white rice.They worked in four mega-varieties of *Oryza sativa* and Indica rice. These Golden rice varieties were planted (following registration by national authorities) in some countries. In others, they will be used as breeding parents to introduce the trait into different locally preferred and agronomically adapted rices.Golden Rice is the first purposefully created biofortified food. It is a rice that synthesises and accumulates β-carotene during seed maturation (Ye et al.2000).The human body efficiently converts the β-carotene in Golden Rice to vitamin A (Tang, 2009). Scientists at Syngenta took other genes which are known to be associated with beta-carotene synthesis in seeds, those from maize, and were able to induce synthesis of higher levels of beta-carotene. Increases in the contents of not only iron and zinc (Trijatmiko et al.2016), but also folate

(Blancquaert, 2015; Paul et al, 2017) and pro-vitamin A in staple crops (Bai, et al.2011; Roberts et al. 2016, Low et al. 2007) are all very important for human health. These have been achieved using GMO-technology, where conventional breeding was not possible. The work with orange-fleshed sweet potato is noteworthy in which the consumption of the crop increased vitamin A intake and serum retinol concentrations in young children in rural Mozambique (Low et al. 2007). Work of Tang et al., (1999) has demonstrated that green and yellow vegetables can maintain body stores of vitamin A in children (Tang et al., 2015). Golden Rice is the most advanced biofortified rice towards a product, and other GM traits for zinc, iron and folate have been worked out (Trijatmiko et al, 2016; Storozhenko et al, 2007; Tang et al, 1999). Transgenesis has also been tried to introduce provitamin A traits into other important food security crops. Beta-carotene, pro-vitamin A, also gives the colour to Golden rice. At the levels found in foods, including Golden rice, beta-carotene is classified as GRAS (Generally Considered as Safe). Elsner et al.(1998) studied the photopigment distribution. There are many studies on the nutritional value of soy and sorghum but the fortification is very low (El Rasheed Ahmed Salim et al., 2017).This could be improved with rural agricultural production, rural technology adaptation, rural processing and fortification. The fortification can be done with wheat or soy bean or other fortification additives. This can improve the health by modifying local foods in rural and urban areas for pregnant women, infants, children and adults. It is based on the ability to formulate nutritive food and their production technology can have food security in rural areas. The methodology depends on the information which is obtained from historical literature and surveys. Fortification provides a nutritionally high balanced food. Locally designed machines for milling, boiling, roasting and packaging by transfer of technology and innovation are being adopted. Some Sudanese diets like fortified cooked sorghum soft sheets (Kisra), fortified thin porridge (Madeda), fortified thick porridge (Logma or Asida) with good protein, fats, minerals and fatty acids are the traditional dishes which can be fortified. This will combat protein-energy malnutrition which is one of the four major forms of malnutrition in developing countries. Soybeans is one of the world's most valuable oil seed legumes which can alleviate protein-based malnutrition and the demand for soybean is relatively high because of high cost of other Sources of protein such as meat, eggs and even cowpeas (Addo and Ogumntona, 1993).Presently, many countries use fortified wheat. Protein in human diets is obtained from several Sources that include cereals, vegetables, legumes, meat, egg, milk and fish (Priya Rampal, 2018).Newer technologies have helped to segregate the readily digestible fractions of foods from non-digestible fraction such as vegetable protein. This has changed the eating habits in many countries (El kalif AEO and Tinay, 2002).Food safety describes handling, preparation and storage of food that can prevent food borne illnesses. Food labeling, food hygiene, food additives and pesticide residues, as well as policies on biotechnology are important on the ready products.Food guidelines for the management of governmental import and export inspection and certification systems for foods which are associated with biotechnology and innovation must be included on ready products.Sorghum and millets are grown in harsh environments where other crops grow or yield poorly. They are grown with limited water resources and usually without application of any fertilizers or other inputs by small-holder farmers in many countries and are known as coarse grain or poor people's crops. They do not find much trade in the international markets and in local markets in many countries. Sorghum is one of the first crops in East Africa (House, 1985). It has been an important crop of semi-arid tropics of Asia and Africa for centuries. Among the sorghum cultivars taken into cultivation the variety, *Sorghum bicolor* (L.) Moench sp. is the most tolerant crop to drought and hot temperatures, diseases, pests, various soil types, has a higher water use efficiency and a higher production capacity from unit area (Chaudhuri et al.1986). Many types of foods include, roasted sorghum grains at milky stage (FAO, 1995), Balila (boiled sorghum grains; Salim, 2016), Kisra (thin sheet breads of sorghum; FAO, 1995), accedah or logma, stuff porridge from sorghum flour (FAO, 1995). Madida or Nasha is a thin porridge from sorghum flour (FAO, 1995), holomore or abrey is a beverage from cooked rooted fermented thick sheets of sorghum flour (Rasheed et al., 2017). Among the sorghum alcoholic beers, bagania is an alcoholic beverage from fermented dry Kisra (Salim, 2016), marisa, is a sour opaque beer which is an alcoholic beverage from fermented sorghum (Salim, 2016) and aragi alcoholic beverage from fermented sorghum flour (Salim, 2016) which are commonly consumed. The important food diets are (Kisra), fortified thin porridge (Madeda) and fortified thick porridge (Logma or Asida) (Salim, 2016; Rooney and Mc Donough, 1987). Gupta et al. (2015) has given an Indian perspective on biofortification of maize.

Wheat

Wheat in Africa is the second major crop after sorghum in terms of consumption. Many wheat products have become common. In one year the country consumes around one million tons (FAO, 2001). Wheat contains 55% of carbohydrates (Breiman and Graur, 1995). The wheat flour contains a large amount of protein and high quality of gluten (Caballero et al.2007). As a food, gluten gives porous characteristic to the bread (Cornell and Hoveling, 1998). The soybean (Glycine max) from family Leguminosae, subfamily Papilionoidae originated in Eastern Asia.It is an industrial crop extensively cultivated for its oil and protein content (Berk, 1992). It contains 40% proteins and 20% oil. Soybeans contain the highest protein content among food crops and second highest oil content among all food legumes.Thus, the composition of soy products range in protein content from about 40% for full-fat flours to 95% or more for protein isolates (Riaz, 1988). The proximate composition of soybean flour and wheat flour showed 3.6% and 9.0% moisture content respectively, 52.5% and 12.9% protein, 12.3% and 1.5% fat, 7.72% and 1.2 ashes, 15.36% and 76.4% carbohydrates, 10.36% and 3.2% fiber respectively (Mohamed et al., 2015).They contain a large quantity of proteins which have an important functional property (Riaz, 1988). Soy flour is often mixed with other flours to increase the protein content and soymilk provides an alternative Source of protein for people allergic to the protein in cow's milk (Wolf and Cowan, 1975).The production of flour from soy bean's involves the removal of hulls to produce a full soy flour or removal of both hull and oil fraction to produce defatted flour that has an excellent nutritive value.The adaptable technology must be feasible, economically viable, sociocultural compatible with the existing farming systems. The agricultural technology transfer should be the function designed to disseminate improved technologies to the end-users (Wolf and Cowan, 1975). For sustainability the farmers should learn the techniques through an informal educational process which would make them self-reliant and make them leaders in their fields. Many studies have shown the possibility of incorporating both; as whole and decorticated grain along with wheat flour at various levels to produce the bread (Adedipe et al.1995) as an enrichment process of adding micronutrients (essential trace elements and vitamins).Nutrient supplementation of foods was mentioned for the first time in the year 400 B.C by the Persian physician Melanpus, who suggested addition of iron filings to wine to increase the potency in soldiers' (Tsen, 1976). Diets that lack variety can be deficient in certain nutrients. Sometimes the staple foods of a region can lack particular nutrients, due to the soil of the region or because of the inherent inadequacy of the normal diet. Addition of micronutrients to staples and condiments can prevent large-scale deficiency diseases in these cases. As defined by the World Health Organization (WHO) and the Food and Agricultural Organization of the United Nations (FAO), fortification refers to the practice of deliberately increasing the content of an essential micronutrient, i.e., vitamins and minerals (including trace elements) in a food irrespective of whether the nutrients were originally in the food before processing or not, so as to improve the nutritional quality of the food supply and to provide a public health benefit with a minimal risk to health, whereas enrichment is defined as synonymous with fortification and refers to the addition of micronutrients to a food.Combination of soaking and roasting are recommended due to economic consideration and reasonable organoleptic acceptability levels (Messina, 1989).According to the FAO, the most common fortified foods are cereals and cereal based products, milk and milk products, fats and oils, accessory food items, tea and other beverages and infant formulas.The methods used for food fortification includes mainly biofortification (breeding crops to increase their nutritional value, which can include both conventional selective breeding and modern genetic modification), synthetic biology (addition of probiotic bacteria to foods), commercial and industrial fortification (flour, rice, oils (common cooking foods) and home fortification (vitamin D drops) (Sanni and Sobamiwa, 1993).Food supplements include vitamins and co-vitamins, essential minerals, essential fatty acids, essential amino acids, phytonutrients and enzymes.Council of Food and Nutrition of the National Academy of Sciences of the United States makes sugar as the criteria for fortification with specific nutrients (Liyanage and Hettiarachchi, 2011; McKingney, 1983). Body builders use amino acid mixtures, vitamins and phytochemicals to enhance natural hormone production, increase muscle and reduce fat. Talsma et al., (2013) studied biofortified cassava with provitamin A to be acceptable by primary school children. Tang (2013) used plant foods rich in beta carotene to combat vitamin A deficiency. Tapsell et al.(2006) studied the health benefits of herbs and spices.

For programmers of fortification in developing countries many factors have been considered like the food should be consumed by all population, the daily intake of the carrier food should be essentially constant, fortification should not alter the organoleptic characteristics of the vehicle, production and processing of the food vehicle should be centralized and the cost of the fortification should be economical and affordable.

There have been concerns over food security from time immemorial. The final report of the 1996 World Food Summit states that food security exists when all people, at all times, have physical and economic access to sufficient, safe and nutritious food to meet their dietary needs and food preferences for an active and healthy life (https://www.ft.com/content/584ed992-4216-11e3-9d3c-00144feabdc0; Chromiak and Antonio, 2002). At the household level it exists when all members, at all times, have access to enough food for an active and healthy life (FAO, 1996). Each person should be food secure and must not live in hunger or fear of starvation (FAO, 2008). According to WHO there are three pillars which determines the food security. It includes food availability, food access and food use (FAO, 2006). According to FAO all these should be sustainable over time (FAO, 2008).

World Summit on Food Security: The Food and Agriculture Organization of the United Nations (FAO) called the summit in response to widespread under-nutrition and growing concern about the capacity of agriculture to meet future food needs. The conference produced two key documents, the

Rome Declaration on World Food Security and the World Food Summit Plan of Action (Chromiak and Antonio, 2002; Bickel et al., 2000). In the World Summit on food security (2009) it was stated that the four pillars of food security as availability, access, utilization and stability of food (FAO, 2013).

The safety between industry and the market and then between the market and the consumer needs attention. Food safety considerations include the origins of food, food labeling, food hygiene, food additives and pesticide residues. Policies on biotechnology, food and guidelines for the management of governmental import and export inspection and certification systems for foods should also be considered. In considering market to consumer practices, the usual thought is that food ought to be safe in the market along with safe preparation and delivery of the food to the consumer. The diseases in food can be transmitted from person to person as well as serve as a growth medium for bacteria that can cause food poisoning. In developed countries there are intricate standards for food preparation, whereas in lesser developed countries the main issue is simply the availability of safe water.Human kind has been utilizing biotechnology in agriculture, food production and medicine. They provide good knowledge for the production of fermented foods incorporated with soybean (Rasheed et al., 2017), beside more development of breeding of soy.In industrial economics, innovations are created and found empirically from services to meet the growing consumer demand (Bhasin, 2012; Ruzzene 2015) www.fda.gov/food/ingredientspackaginglabeling/gras/scogs/ucm261245.htm.

Brenna et al. (2009) found that alpha linolenic acid supplementation gets converted to long chain PUFA in humans. UN department of public information (UNDPI, 2010) has given the fact sheet on millennium development goal. UN educational scientific and Cultural organization (UNESCO, 1998) have discussed on bridging the gap and act for women and girl's education in South Asia.UN (2015) have looked into the levels and trends in child mortality due to malnutrition.UN(2015) UNAIDS have brought forward the fact sheet. UN (millenium development goals, 1990-2015) has also undated the goals which were accessed in 2017.

9

Modern Technological Developments with Respect to Traditional Foods

For Indian foods

Food processing industries are the fastest growing in India. India is the largest producer of milk, bananas, mangoes, guavas, papaya, ginger and okra. It is the second largest producer of wheat, rice, fruits, vegetables, tea, sugarcane and cashew nut. It is the third largest producer of cereals, coconut, lettuce, chicory, nutmeg, mace, cardamom and pepper at the global level (Highlights of Agriculture Census, 2010-11, http://pib.nic.in/newsite/PrintRelease.aspx?relid=132799). Most part of the income is spent on foods. It is important to establish and nurture mutually beneficial relationships with global food processing, food retail and related supply chain organizations. This will in turn help in the realization of significant business growth opportunities in India, through newer technologies, innovations and ways of value additions. The geographical location of India helps in the export of items to other countries. It is connected to Europe, Middle East and Africa from the western coast and to Japan, Singapore, Thailand, Malaysia, Korea, Australia and New Zealand from the eastern coast. Food processing is a priority sector for the Indian Government, especially in recent times with "Make in India" initiative.The country is considered as one of the most favourable markets because of the easy and affordable credit and other fiscal incentives. The total consumption of food and beverages in India is US$ 369 billion which is expected to increase to US$1.142 trillion and the output in the food processing sector might increase to US$ 958 billion by 2025. This offers a vast market opportunity by the Indian food processing, food retail, transport, logistics and related infrastructure sectors to players in the food processing value chain. The market for plant and machinery in the food processing sector is estimated in the year 2024-25 to be USD 51.41 billion as compared to the current value of USD 17.3 billion (2016-17). Leading Global Equipment Manufacturers (GEMs) in India has the ability to address the needs of the food processing sector. Some of the leading GEMs are operating in India. They not only manufacture but make partnerships and joint ventures. This will help the domestic market in an economical manner. The current Indian food processing ecosystem and the impetus given by the Government of India is the right time for the global equipment manufacturing firms to set up a manufacturing base for machinery and equipment. The Government of India, Ministry of Food Processing Industries (MoFPI), is a part of the focus sectors under Make in India.It is driving the sector's growth by creating an environment which can promote for higher investments through sector-friendly policies and grant-in-aid support for the sector. Cereals, oilseeds, pulses, herbs and spices are available on a wide range, and the high level of consumption provides an opportunity for different international businesses which becomes a part of the food industry in India. The contribution of the food processing sector to Gross Value Added (GVA), employment and investments is significant. India is an ideal place for investment in the sector. Many initiatives by the government of India are also providing 100 per cent Foreign Direct Investment (FDI) in food processing, warehousing, storage and transportation for cold chains and marketing of food products produced and manufactured in India. These initiatives give a large number of opportunities to the foreign partners to invest in India.The initiatives of Government of India are taken up by the MoFPI in supporting the food processing sector. All efforts are taken up by the ministry to encourage investments in the sector. Proposals are taken up for joint ventures, foreign collaborations, industrial licenses and 100 per cent export oriented units. The Ministry has also launched the Pradhan Mantri Kisan Sampada Yojna (PMKSY) to supplement agriculture, modernize processing and decrease agri-waste. PMKSY is expected to augment technology infusion and enhance forward and backward linkages between farmer producer companies and the food processing industries. It also helps to develop agro-processing clusters, create a robust supply chain infrastructure and promote skill development in the sector. PMKSY has an allocation of USD 901 million and is expected to leverage investment of USD 4.72 billion. Handling of 334 lakh Metric Tons(MT) agro-produce with a value of USD 15.6 billion will benefit 20 lakh farmers and will generate 530, 500 direct or indirect employment in the country by the year 2019-20 (http://mofpi.nic.in/sites/default/files/databank.pdf).Food processing sector has emerged as one of the major growth-oriented sectors in India due to the demand for food products, abundant supply of raw materials and many incentives given by the government. There are many medium and big food processing industry which import critical components required for their processes. For small scale industries the cost of the imported equipment becomes very high and unaffordable.Therefore, the use of local fabricated machines and equipments is usually made, which are affordable. A lot of R and D needs to be done for good precision. The government is encouraging the technological advancements for the industry by providing specific incentives to the sector in terms of duty reductions and duty exemptions https://www.ibef.org/industry/indian-food-industry.aspx. FAO of UN published on Leveraging small and medium enterprises to improve nutrition (2018). Uzoejinwa et al. (2016) studied the small scale food processing enterprises in Nigeria.

Around the Globe

Ronteltap et al. (2007) have studied the consumer acceptance of technology based food innovations and future of nutrigenomics. For infrastructure creation and packaging at the farm facilities and for the cold storages, newer technologies are required across the value chain. Newer technologies are required in the processing of food products which can retain the nutritional value and infrastructure at port gateways with phyto-sanitary facilitation. There is a requirement of equipments for quality, safety and hygiene which should have high precision. Food which comes in contact with the machinery should be of stainless steel and also be free of cracks and crevices to avoid food particles in them. Food processing equipments must be advanced and sophisticated.The machines used in the industry have similar automation which includes material conveying and positioning, heating, drying and cooling. Regular cleaning and disinfecting, gentle handling and precise control of temperature, pressure, treatment times and other process parameters should be looked into frequently. As a large number of digital sensors, pipes, transmitters, tanks, tubes are used in the processing line they should be able to withstand harsh cleaning agents, ranging from steam and water to alkaline solutions, organic solvents, hypochlorites, iodine compounds and nitric acid.Processing of liquids like soups and oils, requires heating and fractionating of equipment, jacketed vessels, in order to allow steam heating or cold water cooling. Dairy processing involves pasteurization, separation, drying, mixing and cheese-making storage. Clean-in-place (CIP) systems are installed in liquid processes usually dairy for the purpose of cleaning the interior of the equipment without disassembling it. The conversion of meat products into finished goods can be done through grinding, cooking, blending, stuffing and trimming. In vegetable processing it is sorting of color, shucking, mixing, cooking and chilling. Every product has different packaging requirement. For vegetables, machineries are required for different steps like; packaging, canning and freezing. Liquids are canned, boxed and packaging of milk products is decided based on the end product. Meat can be packaged fresh or it can be frozen. The recent packaging equipments have digital controls and operates at a high speed. For meat packaging the procedure is very labor-intensive. In 2016-17, India had imported many food processing equipment from China, Germany and Italy. In the Global players the industries include GEA Group, Tetra Laval, Buhlar, Alfa Laval and Heat and Control.These companies supply various equipments across the value chain for product processing. German manufacturing company, GEA is building a turnkey plant for Amul Fed, the largest milk powder plant in Asia to be fully automatic with the latest technologies.It would produce a capacity of 150 tons per day of skimmed milk powder and 120 tons per day of dairy whitener/baby food. The new plant will process around 90, 000 litres per hour of milk to produce multiple value-added products. Swiss Manufacturer, Buhler Group entered India as Buhler India Pvt. Ltd which is a subsidiary of Buhlar and has a manufacturing plant at Attibele, near Bangalore. It is a leading supplier of equipment for rice milling, flour milling, colour sorting, animal feed technology, oil milling, grinding and dispersion technology and die casting systems. This has a turnover of more than USD 35.6 million. Bühler's Atta process with PesaMil is an industrial process technology, which uses circulation system along with high-compression that enables the production of authentic whole wheat Atta flour. Its taste and texture are identical to Atta flour produced on traditional Chakki stone mills in India. It is also exported to other countries like Pakistan, Nepal, Bangladesh, Burma and Sri Lanka. http://www.gea.com/en/news/corporate-news/2017/gea-to-build-asia-largest-milk-productionfacility-in-india.jsp;http://www.buhlergroup.com/global/en/home.htm#.WdB0sWiCzIU

Manufacturing sector includes; dairy, fruits and vegetable processing, cereal processing, vegetable oil, animal oil and fat processing, meat and meat products, soft drinks, mineral water and other beverages, prepared meals and other food, sugar and sugar based confectionaries including cocoa and animal feed processing.From 2008 to 2015, it was found that the investment in plant and machinery was around 81 per cent of the fixed capital with the gross value of plant and machinery compounded to an annual growth of 11 per cent. However, in the year 2013- 2014, the growth rate in fixed capital was 13 per cent with an output higher by 2.5 times in terms of growth rate as compared to the previous year. Projecting the growth of the segment for the next 10 years at the current growth rate of 13 per cent, the investment in fixed capital is estimated to be USD 9, 815 million by the year 2024-25. Factoring the growth for gross value of plant and machinery at 81 per cent of fixed capital, the expected size of the segment is estimated to be USD 7, 951 million by the year 2024-25. The National Center for Cold Chain Development have studied the capacity assessment of cold chain infrastructure namely, All India Cold-chain Infrastructure, 2015. They found out that there were massive capacity gaps in the cold chain infrastructure in India. Their study highlights the gaps in infrastructure at various levels like modern pack house, reefer transport, ripening chambers and cold chambers. As per the study, maximum gaps exist for preservation of infrastructure immediately after harvest (pack houses; 99.6 per cent gap between requirement and existing) and ripening chambers (85 per cent gap between requirement and existing) followed by transportation infrastructure of reefer transport (91 per cent gap between requirement and existing). Loans to food and agro-based processing units and cold chains have been classified under agriculture activities for PSL as per the revised RBI Guidelines12, issued on 24th March, 2015. The Government has set up a Special Fund of INR 2000 crore (USD 300 million), in National Bank for Agriculture and Rural Development (NABARD) to make low cost credit available to designated food parks and agro processing units in the designated food parks. There is a Special Fund of INR 5000 crores (USD 750 million), to support infrastructure creation for the storage of agricultural commodities (min 5000 MT). Pradhan Mantri Kisan Sampada Yojana (PMKSY) PMKSY scheme has an allocation of INR 6, 000 crore (USD 882 million) for the period 2016-20. It is expected to leverage investment of INR 31, 400 crore (USD 4.71 billion) through this scheme in the food processing sector. PMKSY is an umbrella scheme, incorporating the ongoing schemes and the new schemes of MoFPI to support the sector. It is a comprehensive package to give a renewed thrust to the food processing sector in the country. PMKSY is supporting the ongoing schemes for mega food parks, integrated cold chains and value addition infrastructure, food safety and quality assurance infrastructure.There are several new schemes introduced under PMKSY towards infrastructure for agroprocessing clusters, creation of backward and forward linkages, creation or expansion of food processing and preservation capacities. These schemes are aimed at the development of modern infrastructure to encourage entrepreneurs to set up food processing units based on cluster approach. Furthermore, these schemes provide effective and seamless backward and forward integration for processed food industry by plugging the gaps in supply chain and creating processing and preservation capacities along with the modernization or expansion of existing food processing units. The implementation of PMKSY will result in the creation of modern infrastructure with efficient supply chain management from farm to the retail outlet. In addition to the various schemes of MoFPI, there are other agencies, which provide a number of grants to the sector.http://pib.nic.in/newsite/PrintRelease.aspx?relid=16151214;http://www.mofpi.nic.in/sites/default/files/important_notice-sampada-19.05.2017.pdf

Tetra Laval Group consists of three industry groups, Tetra Pak, Sidel and DeLaval. All the technologies provide efficient production, packaging and distribution of food. DeLaval is more focused on dairy farming and animal husbandry and Tetra Pak on processing and packaging technologies. Globally, Tetra Pak operates in more than 175 countries. In India, Tetra Laval, operates through, Tetra Pak India Pvt. Ltd. It has two major domains; processing and packaging for application processing of dairy processing, beverage processing, prepared food processing, ice cream processing along with cheese and whey processing. It is a pioneer of packaging and key work domains are manufacturing of filling machines, downstream equipment and automation. http://www.tetrapak.com/in

SSP is a process engineering company that offers equipment and process solutions for food, dairy, beverages and chemical Industries.The company was established in 1977. Major product categories of SSP are dairy industry, instant coffee and tea, starch and derivatives, fruit and vegetables, nutraceuticals and protein food, edible and pharma grade salt, dairy alternatives, egg/ honey processing plant. http://www.sspindia.com/about-us.html

Atlas Pacific Engineering Company, Inc. which was established in 1945, is a global leader in the fruit processing industry. It offers all types of fruit processing machinery.They work under three different businesses, Magnuson corporation, Luthi machinery company and Brown international corporation. Each company operates for different product segment. Magnuson product lines include peelers, feeders and corn processing equipments which are computerized. Luthi Machinery has can filling technology for the tuna industry, offering machines for solid pack, chunks and flakes. It specializes in sold filling equipment for the chunk and solid pack. Brown's citrus system typically include extractors, finishers, by-product recovery systems and plant processing control and information systems. It has non-citrus processing units like extractors, finishers, pulpers, classifiers, separators, de-waters and refiners. There is no manufacturing plant in India. Most of the machinery in India are imported. https://www.atlaspacific.com/ Bucher company supplies agricultural machines that deal with soil cultivation, sowing technology, fertilization, plant protection, landscape maintenance, animal feed and feeding technology. http://www.bucherindustries.com/en/divisions/bucher-specials;http://www.bucherindustries.com/en/about-us/locations

The world's leading manufacturers of food processing, seasoning, conveying, packaging and inspection equipment systems is Heat and Control. The company in India is known as Heat & Control South Asia Private Limited. They manufacture a wide range of equipment across the value chain of food processing and provide complete solutions in the areas of process prepared systems, packaging systems, snack systems, french fry systems, product handling systems, inspection systems and control automation system. http://www.heatandcontrol.com/contact.asp

In 1988, a food processing and packaging company, namely Bajaj Process Pack was established which manufactures and provides turnkey solutions for various equipments utilizing fruits, vegetables and beverages for small and medium level processing units. It provides equipment for processing of fruits and vegetables which includes; pulp, puree, paste, juice concentrates, jams, juices, ketchup, dehydrated fruits and vegetables and end-packaging in cans, bottles and pouches. http://www.bajajmachines.com/

A company called Economode Food Equipment was established in the year 2001, and manufactures automatic machines. It provides fully automatic processing lines for potato chips and fully automatic nik-nak (corn and rice puff) and namkeen /ethnic snacks production lines. They have both peak performing continuous frying systems and batch frying systems and extruders like cheese ball extruders and kurkure extruders. http://www.economode.in/

Mumbai based, Mazda group is a manufacturer of food and beverage processing plants. It provides solutions for design, fabrication manufacture and installation. It has specialized manufacturing facilities in India and in Europe. The specializations are in various products as that of french fries / potato products plants, tomato paste and ketchup plants. They have fruit extractions plants for juices, purees, pulps, frozen fruits and vegetables and cold chain elements.They also have aseptic packaging and storage for meats, sea foods and RTE Food. http://www.mazmach.com/index.html

Satake Corporation, Japan has a subsidiary company known as Satake India Engineering. In India it started its operation in the year 1996. It produces individual machines, integrated systems and engineered solutions for the processing of rice, wheat and other grains. It works in the designing, supply, installation and commissioning of machinery for food industries for the crops like rice, wheat flour, maize and lentils.They have commissioned more than 150 modern rice milling plants and over 700 color sorters. http://www.satakeindia.com/

In Taiwan there are many food machine industries. Among them, ANKO Food Machine Co., Ltd. is a leading company which offers a wide range of food equipment. Globally, they supply their machines to more than 109 countries. They have machineries for frozen food machines to specialized bakery machineries that include for puff pastry, mammoul, burrito, bliny, apple pie, cinnamon roll, machines for seafood and meat value added products. Apart from individual machines, they also take up turnkey projects starting from factory planning, raw material processing, production line planning, production process optimization, new product development and also customization. In India, they have provided automatic machineries for a large number of Indian food products ranging from lachchha parathas to samosas, dumplings, namkeen and peda. https://www.anko.com.tw/en/country/IN.html

Sifter International was established in 1978. Starting with manufacturing of the cleaning and grading equipment for grain processing industry.Sifter is manufacturing equipment across a range of products.Sifter International deals in food processing machines, turnkey project solutions food processing plants, food packaging plants and machineries as well as exports the equipment and turnkey solutions. It is executing project for flour mill, rice mill, spice plant, cleaning and grading plant, solvent extraction plant, maize mill, cornflakes plant, tomato processing plant, different industrial dryer, guar gum plant, fruit and vegetable processing plant and dehauling of seeds. http://www.sifterinternational.com/index.php

E & S Food Technologies Pvt. Ltd. was established in the year 2012 which manufactures a wide range of food processing machines, storage silos, spiral conveyor, pasta machines and snack dryers, cookies making machine, material handling station system, food processing machinery, snacks stamping machine, pasta machine, spiral conveyor, stuffing machine, macaroni making machine, storage silos, dryer machine, automatic pasta lines machine, pasta machine die and inserts.http://www.esfoodtechnologies.com; http://www.esfoodtech.com/

Indopol Food Processing Machinery and manufactures is an independent family-run business group which started in 1983.It provides various types of cereal milling machinery and accessories, ready-to-eat energy foods as well as provides technical consulting for food processing both for import and export purposes. The specialization of the company is in flour and rice milling. The products include; roller flour mills, hi-tech rice cleaning, parboiling and drying, milling and grading plants, blended cereal foods and food extrusion systems, cleaning of pulses, grading and milling plants, grinding of spices and grading of plants, other food processing machinery solutions, automation and systems controls for food processing systems.http://www.indopol.com; https://www.zaubacorp.com

Meyn Food Processing Technology B.V is a company which is a subsidiary of CTB Inc. It is a leading global designer, manufacturer and market systems and solutions for preserving grain; producing poultry, pigs and eggs; processing poultry for various equestrian and industrial applications. The company was started in the year 1952. Since 2002 it became as Berkshire Hathaway company. They have a global network and work through a worldwide network of independent distributors and dealers. https://www.meyn.com/

Services to the fish, meat and poultry industries are provided by a global leader provider of advanced equipment and systems called as Marel. Later in 2012, Marel India was established as a part of the Marel global sales and service network. It operates worldwide in over 30 countries and 6 continents. The main office for Marel India is located in Bangalore. They have service departments in Bangalore and on the west and east coasts of India. http://marel.in/

Jarvis Products Corporation are the manufacturers of specialized machinery that processes meat especially meat cutting and boning machinery. It also includes band saws, slaughter equipment, carcass opening and splitting saws, hock cutters and skinning machines as well as other power tools for the poultry and fish processing industries. Jarvis equipment is sold and serviced worldwide through different branch locations. The Jarvis' J26 Federation of distributors, Jarvis Equipment Pvt Ltd, India is completely owned and is a subsidiary of Jarvis products corp., located in Delhi, NCR region. This is a single point of contact for their customers in India and neighbouring countries dealing in sales and service of Jarvis tools.http://www.jarvisproducts.com/; https://www.jarvis.co.in/

RND Automation Pvt. Ltd is a company which provides help in designing, manufacture and customizing poultry and meat processing equipment.The company manufactures from semi-automatic defeathering plants to automatic conveyorized plants for processing between 300 birds/hour to 4000 birds/hour, utilized by both small scale and large scale processors and integrators.The company is globally having units in Belgium, Bangladesh, Sri Lanka, Abu Dhabi, Nepal, Libya, Nigeria, Sudan, Kenya, Surinam, and Myanmar. http://www.jarvisproducts.com/ ; https://www.jarvis.co.in/

Frick India is a company which gives its services in manufacturing, package, installation and tailor-made refrigeration systems for quick freezing and low temperature storage of meat and poultry industry and sea food products. It has fully automated refrigeration solutions like quick chilling rooms, water chillers, pre-chillers, blast freezers (up to -50°C), plate freezers, low temperature frozen chambers and cold storage, chilled rooms, water chillers, pre chillers, IQF, flake ice machine, low temperature frozen chambers, ice production and storage. http://www.frickweb.com/meat.aspx

John Bean Technologies Corporation (JBT) is a leading global technology solution provider for high-value segments of the food processing and air transportation industries. It provides technologies for freezing and chilling of meat, seafood, poultry, ready-to-eat meals, fruit, vegetable and bakery products, protein-processing solutions that prepare, portion, coat and cook poultry, meat, seafood, vegetable and bakery products, shelf-stable sterilization solutions for fruits, vegetables, soups, sauces, dairy and pet food products, as well as ready-to-eat meals in a wide variety of modern packages.Fruit and juice processing solutions that extract, concentrate and aseptically process citrus, tomato and other fruits. It gives High Pressure Processing (HPP) solutions for ready-to-eat and ready-to-cook meats, ready-meals, fruits and vegetables, juices and smoothies, soups and sauces, wet salads and dips, dairy products, seafood and shellfish. JBT FoodTech Pune, Indian sub company of JBT is mainly into the processing of citrus, other fruit and vegetable processing, aseptic filling, closing, sterilization, freezing, proofing, refrigeration, portioning, slicing, coating, frying and filtration, cooking, process, line control and pasteurization. http://www.jbtfoodtech.com/

Rheon Automatic Machinery provides automatic production of traditional foods from around the world. Rheon has been developing, manufacturing and supplying a wide range of high quality food processing machines and factory systems. They are the world's first developer of fully automated croissant production line and automatic filled product machine. They export to over 120 nations. www.rheon.com/index.php

Kanchan Metals Pvt Ltd is a private company started in1984. It manufactures equipment and provides solutions to food processing companies in India.The food processing equipments are of three types provided by the Kanchan Metals. These are continuous namkeen extruders, belt conveyors, forced air coolers, seasoning drum and bucket elevator. Their major equipment work includes snack food manufacturing equipments, meat processing, bakery, confectionery, wafers and instant foods, packaging and hygiene. http://snackfoodmachines.com/

Sanjivan Industries established in 1996, is a food processing products manufacturer and supplier. It gives a range of sweet making machines, kalakand kadai, boiling kettles, namkeen making machines, namkeen plant, food processing machines and dal frying kadai. http://www.sanjivanindustries.com

LTX-Credence Corporation has a subsidiary company known as Dover India Pvt. Ltd. It designs and manufactures various components for communications, life sciences, aerospace/industrial and defence markets.It provides engineered solutions for the extraction and handling of critical fluids for drilling, production and for the downstreaming of oil and gas markets, including power generation worldwide. It has systems for fluids, refrigeration, food equipment, waste and recycling and industrial markets. Further, it provides integrated printing, coding and testing solutions for consumer goods, food, pharmaceutical, industrial, electronics and alternative energy markets. http://www.sanjivanindustries.com

Under HRS Group there is a unit called as HRS process systems limited.It operates for thermal processing technology and is one of the Europe's leading heat exchanger manufacturer for multiple sectors. In the food processing industry it has equipment for design, manufacture and installation of complete heat treatment systems for the food industry for aseptic filling, hot filling or any other form of filling. For the food processing industries they also have integration of pumps, tanks, heat exchangers, aseptic fillers, piping, and controls etc. https://www.hrsasia.co.in/

It was found necessary for many products to be refrigerated. Industrial Refrigeration (IR) was founded in 1958. IR started progressing in food refrigeration post 1990's and over the years gained experience in ice making machines, freezing plants, chillers and cold storages for temperatures in the range of + 5°C to - 40°C and for applications ranging from storage of films to storage of seafood. During these years IR has entered into many tie-ups with global companies like GEA group to introduce the latest refrigeration technologies and practices in India like paraffin wax freezers for refineries, snow rooms, ice skating rinks, dairy refrigeration, abattoir refrigeration and modified and controlled atmosphere cold stores. http://www.irl.co.in/.

Innovation processes involve in identifying the customer needs, macro and meso-trends, developing competences and finding financial support.The agricultural scientists are trained to manage agricultural innovation as a process of vertical transfer. Processes should be helpful to the farmers and that the ideas and needs can be directly involved with scientists in developing food recipes and prototype technologies.It is the policymakers who have worked to develop environments to foster innovation.

Newer machines house hold hand milling, small scale soyabean machine, hammer mill and sorghum mill are available in the market by various manufacturers.

Integration of innovation and technology transfer in sorghum food security fortification

In order to improve the nutritional value of sorghum it can be supplemented with soybean which provides energy, protein, fat and weight (Peter, 1985).

Processing of fortification components

Dough: Dough has been described as the most complicated rheological material because the properties of dough are connected with baking characteristics of flour. The desirable rheological characteristics of dough (stickiness, mixing requirement, water absorbance) together with the color and appearance of loaf are of great importance (Muller, 1969).

Milling techniques: Different milling processes (hand pounding, roller milling and abrasive decortication-hammer milling) are utilized most of the times for milling of grains (Finney et al.1987).

Small-scale service and/or semi-commercial milling: RIIC/PRL type dehuller is very suitable for most sorghum processors. It is also cost effective and has low maintenance costs, than other available decorticators like the peritech (Finney et al.1987).

Suggested improvements for Soy-sorghum milling processes: The biggest challenges in sorghum milling are associated with the difficulty to remove the pericarp and the germ from the endosperm (Bassey and Schmidt, 1989).

Treatment of anti-nutritional compounds: There are many anti-nutritional compounds like tannins, phytohaemagglutinins, protease inhibitor, estrogenic factors, phytic acid, dihydroxyphenylalanine, amylase inhibitor, cyanogen's, phytohaemagglutinins, flatulence factor and glucosinolates found in oil seeds and pulses (Tacon, 1997). They are removed by heat treatment, autoclaving (121°C:15-30 min), fermentation, roasting at 100°C or higher, steam autoclaving (120°C:2 h), micronizing, aqueous heat treatment, (100°C:10 min), solvent extraction with hexane, aqueous extraction (18 h), soaking in water, microwaving and soaking in copper sulfate solution. All these innovations and fortifications have improved the physical and nutritional values achieving food security and food safety for nutritive and healthy food.

Innovative Agro Business Models: Some of the innovative agribusiness models are based on coconut virgin oil, sugarcane juice, high curcumin in turmeric, disease free banana, solar dried fruits and vegetables, nutrient bar, minimally processed vegetables packaging by modified atmosphere, promotion of new crops for healthier India with chia and quinoa, spice mixes for traditional foods, traditional snack foods, spice processing in rural sector and leaf plate machine which can be taken up by small farmers using economical small machines where ever required.

10

Challenges in Traditional Foods and Nutritional Security: Female Education

It has been found that at the household level the health of female members in any family suffers to a large extent. It could be encountered to low health care, poor housing or/and environmental conditions. Due to malnutrition it affects the health to such an extent that it gives rise to many diseases. For a sustainable nutritional status food and various aspects of health needs attention. Looking into that, nutrition is directly related to the health status. For food and nutrition security it is vital to see the availability, accessibility and utilization of food. The World Food Programme (WFP) developed the Vulnerability Analysis and Mapping (VAM) project to analyse the vulnerability to food insecurity of target populations. A prominent part of VAM is related to food access. The demographic and Health Survey (DHS), funded by USAID, provides health data for many countries to help them design their national policy. FAO has developed the Global Information Early Warning System (GIEWS), which collects data related to temporary food insecurity. Under the leadership of WHO, several health surveillance systems have been developed and implemented to monitor the epidemiology of various forms of malnutrition and of selected diseases. At the meso or sub-national level, food market surveys provide data on the availability of food. Agricultural production surveys, intra-household food frequency interviews, immunization surveys and anthropometric surveys of children under five can be used to assess the availability, accessibility and utilization of food and its stability at micro level. Most common Food and Nutrition Security (FNS) indicators at different social and administrative levels depends on the supply and demand. To determine the national situation of food availability data on the production of different food commodities, fertility rate and the trends in internal population needs to be seen. Food prices and per capita food consumption are indicators for national food accessibility. The rates of stunting, wasting and underweight in children, low Body Mass Index (BMI) in adults and low birth weight are FNS impact indicators that designate the extent to which food is adequately being used, utilized and converted into a satisfactory national nutrition situation. Fluctuations and shortages of food availability or accessibility in food prices make the national FNS unstable. Sensitive indicators for accessibility of food are the prices in the market. The sewage disposal and diarrheal diseases (DD) rates affect food utilization.A temporal food and nutrition insecurity arises due to pre and post harvest food availability and accessibility along with low BMI.If the food is in shortage leading to the consumption of more wild foods indicates reduced food availability at household level. Also if the number of meals per day goes down and unemployment increases it indicates low food accessibility.Malnutrition indicators like wasting, goitre or anaemia among the household members show reduced food utilization at micro level. Finally, changes in pre-harvest food consumption practices are good indicators for temporal food insecurity.

Food and Nutrition Security is linked to many issues in the development of individuals and also societies. These include gender, livelihood, poverty reduction, HIV and natural disasters. Women play an important role in FNS as producers of food and as managers of natural reSources, in income generation and in providing care to their families. Inspite of this women have a limited access to land, education, credit, information, technology and in decision making bodies. Women are thus impaired in fulfilling their potential socioeconomic roles in food and nutrition security and in ensuring care, health and hygiene for themselves and their families. Iron deficiency and undernourishment during pregnancy and lactation makes them nutritionally insecure. Women agriculturists are not secured of their own land as they have not protected their rights. Sometimes they get affected due to the changing socio-economic conditions, land shortages and tilting programmes which do not recognize the contributions of women to agriculture. Improving access to land for women is necessary for food security and sustainable production. Other problems for women are fewer employment opportunities, lower wages less access to resources and information, less involvement in decision making lower enrollment at school and earlier drop outs. In many countries, girls are married early during the adolescent years and thereafter start bearing children. Due to gender inequality, women and girls are poorly nourished throughout their life. This gives higher mortality rates, low health care and higher FN insecurity. In many parts of the world, more women are affected by hunger (FAO, 2012). Due to the males being favoured in quantity and quality for land allocation women are insecured. Access to financial services is limited with inequities regarding access to livestock, inputs such as seeds and fertilizers, technology, market information, knowledge, skills and advisory services. Female illiteracy further aggravates the situation (FAO, 2011). The number of people facing hunger worldwide could be reduced upto 17%, if women could access productive resources (FAO, 2011). Women must find the financial means to buy market food. Yet many women in developing countries lack control over the household income, as well as income generation possibilities. Many a times, women give their earnings to their husbands who do not spend it to buy good food. It is therefore, necessary to have an awareness program for men, for FNS and another program for the complete family. It is possible that if a women has a better control on household income, she may use it for better FNS. Tis in turn would benefit children's health, nutrition and education. Amount fixed for food allocation in a household could be a limiting factor. The multiple roles of women are challenging in terms of time allocation. If the workload for a woman increases due to agricultural activities then they are not left with adequate time for taking care of the children. Children born to malnourished mothers often are underweight and face a 20% higher risk of dying below the age of five (UNICEF, 2007). The good health and nutrition of mothers helps in the well-being of their children. An insufficient supply of nutrients to the mother, during a child's first 1000 days can affect a child for the complete life with chronic health problems, cognitive and physical deficits or impaired immune functions. A woman can give birth to well-nourished babies if her nutritional status is sound. Promoting gender equality has been a fundamental principle of german development policy for many years (Alkire et al. 2013). Many steps have been taken and implemented by Deutsche Gesellschaft fur Internationale Zusammenarbeit (GIZ) GmbH on behalf of German Federal Ministry for Economic Cooperation and Development (BMZ). In case the policy can be carried out at all levels, then it would improve the food security. Matter of particular importance is the participation of female household heads and married women in local councils. There must be a better coordination between sectors of agriculture, health and education both at national and regional levels. There should be equal access to land both in quality and quantity.The distribution of land titles and certificates should be gender equitable.There should be equal access to institutions, capital, income, credits and financial advisory services. Rights for (re-) investments and control over income needs to be strengthened. Department of economic and social affairs office (2008) coordinated to achieve sustainable development goals.

Women and girls are overrepresented among those who are food-insecure. Worldwide, an estimated 60% of undernourished people are women or girls (United Nations Economic and Social Council (ECOSOC, 2007). Therefore, efforts must be taken against gender discrimination for access to food as it is a global problem and needs attention. By investing more in women using an approach that is empowering and enlarges their freedoms, it will help the society at large. Solutions to the multiple crises account to the specific situation of women and girls.which needs to be addressed (Gender equality and food security, women's empowerment as a tool against hunger, 2013). FAO and FHI 360(2016) have documented the minimum dietary diversity for women. Food security and hunger eradication are among the top priorities on the international agenda today in view of the impact on agricultural productivity of global economic crises, food price spikes and climate change. The extent to which gender inequalities and the gender gaps in agriculture can be tackled are the priority goals.Women in household and community play a vital role towards food and nutrition security. Across the Asia and Pacific region, rural women assume critical roles in attaining each of the pillars of food security which include; availability, access and utilization. Their role is thus crucial throughout the agricultural value chain, from production on the family plot, to food preparation, to distribution within the household. However, their roles are generally undervalued and constrained by limitations on their access to reSources, services and labor market opportunities. Most rural households and communities in the region manage their agricultural production systems based on social norms and practices that determine the gender division of labor. These vary from the notably constrained roles and opportunities of women in South Asia (with the exception of Sri Lanka) to the more complementary gender roles found in parts of Southeast Asia and some of the Pacific island nations. A lack of data, combined with gender-biased perceptions, have limited awareness and appreciation of rural women's productive roles and contributions to food security in the region. In addition, poor education and health indicators for women limits their access to resources and opportunities. It also constrain the role of women in agriculture and food production. This is inhibiting to bring food and nutrition security in the Asia and Pacific region. Therefore, it is important to bring in priority interventions that would enhance food and nutrition security in the region by ending gender discrimination and empowering women. It argues for policy reforms to advance gender equality and strengthen country-owned food security strategies. Social protection policies that address the structural causes of poverty and vulnerability are strongly recommended, as are policies that improve women's access to agricultural production resources, services and to decent wage employment. Equality of treatment between women and men and food security are mutually supportive, gender equality remains an elusive goal in many regions and a transformation in the traditional gender roles is urgently needed. Such a transformation can be enhanced with improved information about the range of inequalities and specific constraints facing women. A simultaneous and integrated pursuit of such information and transformation is essential for gender equality strategies and food security strategies to complement each other and maximize their synergy. Measures need to be taken to relieve women of their burdens and recognize their largely undervalued contributions to household chores. Such measures must be linked with strategies that pave the way to transformed, equitable gender systems. Women and girls worldwide face many inequities and constraints often embedded in norms and practices and encoded in legal provisions. Some laws, that govern access to land, include inequitable and exclusionary provisions which institutionalize discrimination and needs amendment as it limits the access to key reSources such as land, credit and affects the household food security and nutrition. Not only are women and girls affected directly, but members of their households and communities get affected inter- and intra-generationally. Women and girls are affected through two main channels. One is the limit on their access to education and employment opportunities, which curtails their economic autonomy and weakens their bargaining position within the family. Their weakened bargaining position translates into little or no voice in household decisions, differential feeding and caregiving practices favoring boys and men. It also makes it more difficult for women to fulfill their vital roles in food production, preparation, processing, distribution and marketing activities. Challenging the constraints women face must therefore be treated as a key component in the fight against hunger and malnutrition. Such an approach is achievable, inexpensive and it can be highly effective. A greater role for women in decision making at all levels, including the household, local communities and national parliaments is vital in all respects. Women's empowerment is not only a priority goal in itself but an intrinsic human right. Their autonomy as food producers is usually very limited by the significant obstacles they face in accessing land, financial services, extension services, markets and in benefiting from agricultural research and development. Removal of these obstacles through gender-sensitive approaches would result in significant productivity gains benefiting the households, communities and society as a whole. Implementation of social programs can be used as opportunities to empower beneficiaries and to accommodate the specific needs of women. Social development measures can also be delivered in the course of social program implementation.Provision of child-care services and the redistribution of power within the household are important not only to allow women to make the choices and provide the care that matter for infants, but also to ensure that men value and contribute to such care. Gradual redistribution of household responsibilities must be a central component of any gender-sensitive employment policy. Support of country-owned food security strategies that would maximize the synergy between gender equality and food security across sectors will be helpful. There should be an approach based on the rights which could improve the accountability and encourage independent monitoring of progress. National strategies should be developed in a participatory way, involving women and men. Men should be co-opted into strategies aimed at improving the situation of women to increase their chances of success. Strategies should be in a phased manner and in transformative roles. Any strategy must incorporate a time frame sufficient to undertake the challenges.

Investments are needed in women's and girls' education, health, and economic empowerment, as well as, in public transport facilities, better water services and clean energy sources especially in rural areas. This will help in strengthening the social protection of women that will integrate gender dimensions in their design and implementation. Emphasis should be given to support women for investment in agriculture through technology transfer, through enabling infrastructure like irrigation systems, farm-to-market roads and vendor markets, through measures that benefit women's cooperatives and other associations. Challenging the constraints women and girls face and improving the opportunities is important in the fight against hunger and malnutrition. It is achievable, cost-effective and may form the basis for a sustainable strategy for reducing food insecurity. However, removal of these constraints requires more than good intentions and legislative reform of discriminatory legal provisions. Social, cultural norms and the gender roles that these norms impose must also be challenged. FAO of UN (2017) published the strengthening sector policies for better food security (2017).

The global food price crisis exploded in 2007–2008, placing food security in sharp focus and leading to a succession of international summits. From November 2007 to June 2008, prices of the main commodities went through significant increases on global markets. The food price index rose by nearly 40%, compared to 9% in 2006 (von Braun 2008). Price increases were significant for wheat, maize and rice. The price for maize almost tripled from 2000 to 2008, and the price of wheat quadrupled during the same period, largely due to low stocks resulting from increased demand for agrofuels (Organisation for Economic Co-operation and Development (OECD, 2008).The price of rice also quadrupled from January to April 2008, in large part due to export restrictions imposed by major exporters in what is a relatively thin global market. According to some measures, these increases meant that real prices of food commodities reached a 30-year high in June 2008. Undernutrition and malnutrition of women affect their health and their children. Poor nutrition of a mother during pregnancy and the child during its first 2 years of life has lifelong consequences for the child's physical and mental development (Hoddinott et al.2008). A survey covering the period 1993–2005 found that 42% of pregnant women worldwide suffered from anemia (iron deficiency) which can cause low birth weight (Harvey et al. 2016). In 2008, maternal undernutrition (where a mother has a body mass index of less than 18.5 kg per square meter) affected more than one-fifth of women in south central and southeastern Asia, and more than two-fifths of women in India and Bangladesh. Maternal undernutrition is associated with intra uterine growth restriction, which could lead to the retardation of mental and physical development in the offsprings (Salam et al. 2014). The ongoing food price crisis, financial and economic crisis, could amplify such developments. The food price crisis gives a negative impact on human development by increasing poverty and inequality, worsening nutrition as families shift to diets which are less diverse and have lower micronutrient content. It also reduces the utilization of education and health services and depletes the productive assets of the poor. Deterioration in any of these areas is difficult to reverse and may have ramifications lasting years and in some times even generations (Martorell, 2017; Christian, 2010). Access to food can be achieved by own production, for those who have access to land and can exploit it, employment and self-employment, generating income and thus allowing purchase of food and social transfers including food-for-work or cash-for-work programs and cash transfers or other forms of solidarity within households or communities. Women can face discrimination in all these areas. Their greater vulnerability to food insecurity in times of crisis, compounds the problems of food insecurity. The ability to access food depends on power to produce, power to purchase, and power to access food in intra-household allocation mechanisms. When women have less power than men, it translates directly into weaker access to food (Patel, 2012). The current ecological crisis and especially, climate change impacts are foreseen affecting women disproportionately.The prevailing mechanisms of intra-household allocation are expected to result in increasing undernourishment of women and girls as food prices go up. Additionally, given their predominance as small-holder agricultural producers, women will be particularly hit hard by climate change due to their lower bargaining power and discriminatory practices in the agriculture sector. Declining yields will diminish their ability to feed their families or will result in loss of purchasing power particularly for the poorest rural households who are net food buyers. Many of these households survive by combining food production for their own consumption with income-generating activities. Climate change will reduce food security, not only because of the reduced availability of productive resources and its affect on food supplies, but also because of expected reductions in time for caring practices (Tirado et al., 2011). The anticipated increase in burdens on women and girls implies towards reducing their capacity to exploit opportunities for income-generating activities or education (Tirado et al., 2011; Masika 2002). Researchers from the International Food Policy Research Institute (IFPRI) reviewed existing literature on the use of fertilizer, seed varieties, tools and pesticides. Peterman et al. (2010) found that 79% of the studies concluded that men have higher means to access the inputs. One study in Burkina Faso found productivity on female-managed plots to be 30% lower than the male-managed plots within the same household because labor and fertilizer were more intensively applied on men's plots (Udry 1996). Although, the literature shows equal access to inputs for men and women but in reality it does not happen (Udry et al., 2005).

FAO concluded that if women had the same access to productive resources as the men, they could increase yields on their farms by 20–30 percent. This could raise total agricultural output in developing countries by 2.5–4 percent, which could in turn reduce the number of hungry people in the world by 12–17 percent (FAO, 2010. Globally, women represent two-thirds of the 600 million poor livestock keepers in the world (FAO, 2010). Their role is particularly important in a number of Asian intensive livestock systems, where it is estimated that more than three quarters of livestock-related tasks like feeding, animal care and milking are performed by women (Tipilda and Kristjanson, 2009). In India, women account for 55% of livestock farming labor, while their participation in work related to the care of animals is above 77% and they represent 93% of people employed in the dairy subsector (Richter, 1997).

Teufel et al. (2019) have described the position of women in relation to FNS in Tanzania. They have made a study using mixed-methods.Qualitative research has helped to study the gender dynamics and has found that women's control over assets and income has been associated with dietary diversity to purchase more nutritious foods. This finding corroborates prior research in Ghana (Amugsi et al.2016; Malapit and Quisumbing, 2015a), Ethiopia (Yimer and Tadesse, 2015) and Nepal (Malapit et al.2015 b). To increase the milk production many interventions are required in order to support women which have been done in Mvomero and Njombe districts in Tanzania (Mwaseba and Kaarhus, 2015). However, quantitatively no correlation between women's overall empowerment and household food security has been recorded. But, this finding is not consistent with research from Ghana (Malapit and Quisumbing, 2015a), Bangladesh (Sraboni et al., 2014) and Nigeria (Olumakaiye and Ajayi, 2006). The research describes the influence of various domains of women's empowerment on nutrition and food security outcomes. This could be context specific which is affected by characteristics of the local socio-cultural or agricultural systems (Amugsi et al. 2016; Johnston et al. 2015; Malapit and Quisumbing, 2015a; Olumakaiye and Ajayi, 2006; Sraboni et al. 2014; Yimer and Tadesse, 2015). The authors interpreted in terms of aspirational versus actual empowerment in relation to nutrition. There has been an association between women's empowerment and dietary diversity but no association with food and nutrition security. The quantitative results have shown that men are more associated with food security, whereas women with nutrition security.

The qualitative findings also showed men are considered to be the main decision-makers. Working in depth would provide quantitative studies. Under the DAC Network on Gender Equality (DAC Network on Gender Equality, Justina Demetriades, 2009) it has been looked into the overall limited suitability of quantitative tools to capture gender dynamics. Kabeer (1999) has discussed the importance of opportunities that provide a reflexive vantage point to evaluate relationships (family or community) and reshape individuals' perceptions of themselves that contribute to empowerment. Galiè et al. (2018) have discussed some of the pros and cons of universal versus local conceptualizations and measures of empowerment. It recommends the adoption of a conceptual framework of empowerment that includes economic domains and to capture the link between nutrition and food security.

Notably, the qualitative findings also revealed how a given intervention aims to strengthen the women's empowerment and enhances the food security and nutrition through forage. It is important to focus first on governance issues rather than on the development. Introduction of forage technologies (new crops/varieties) should be studied. Mixed-method approaches are often recommended but very rarely studies are done in detail to address the non-alignment of findings that may arise. In the Ethiopian highlands > 50 % of the population (approximately 20 million people) live under the threat of food insecurity. The government is working towards the betterment. The national strategies are towards the rural development with the cooperation of local agricultural authorities. For the small scale farm householding, one fifth must be female heads and support must be given for sustainable land management. Following these till date, 77, 000 hectares of land have been rehabilitated and another 79, 000 hectares of forest are being maintained by participatory forest management principles. With newer cultivation techniques and erosion control crop and livestock productivity, it has been found to increase the yields both for male and female farmers. Women are benefitted by gaining better access to advisory services, information and skills. Female and gender-sensitive male advisors are helpful. The Ministry of Agriculture has given support to women to access the innovative techniques in watershed. The National Women's Office, supports the establishment of women's groups and the representation of women at community level. Women's groups serve as dialogue platforms for their development priorities, which include the promotion of animal husbandry or fruit/vegetable production creating an additional income. The linkage of sustainable land management methods with HIV/AIDS education and family planning would reduce the pressure on natural resources and increase both food and nutrition security. Equal access should be given to agricultural inputs and advisory services both for livestock and crop production. Advisory guidance must be given along with the agricultural inputs like seeds, fertilizer, pesticides and veterinary medicine which will increase food production. Advisory services must account the financial and intellectual capacity, as well as time availability of the clients and must focus both on men and women. It was observed that gender separated training were more successful which had more open discussions. For capacity development it is suggested to have participatory methods in all phases of project planning and implementation. Support to women for any advisory services and gender sensitization should be given. This will strengthen and create a link between different institutions and stakeholders on gender issues. Project activities should aim to benefit men and women equally.

In order to attain Food and Nutrition security all over the world a few millenium development goals have been set by the FAO. Goal 1: Eradicate extreme poverty and hunger: Food insecurity and malnutrition erode human capital, reduce resilience to shocks and reduce productivity (impaired physical and mental capacity). Goal 2: Achieve universal primary education: malnutrition reduces mental capacity.Malnourished children are less likely to enroll in school. Hunger and malnutrition reduces school performance. Goal 3: Promote gender equality and empower women: Food secure and better-nourished girls are more likely to stay in school and, subsequently, have more control over future choices. Goal 4: Reduce child mortality: malnutrition is directly or indirectly associated with more than 50% of child mortality. Malnutrition is the main contributor to the burden of disease in the developing world. Goal 5: Improve maternal health: Maternal health is compromised by an antifemale bias in allocations of food, health and care. Food insecurity and malnutrition are associated with most major risk factors for maternal mortality. Goal 6: Combat HIV/AIDS, malaria and other diseases: Food insecurity spurs coping mechanisms, such as migratory labor and/or prostitution, which increases the spread of HIV/Aids. Malnutrition hastens the onset of aids among HIV positive. Malnutrition weakens resistance to infections and reduces the chances of survival for those who have malaria. Goal 7: Ensure environmental sustainability: Food insecurity leads to unsustainable use of forestlands and reSources (UN SCN 2004 and FAO 2002). Indicators for food and nutrition security at national and regional level are important. In order to assess and analyse the food and nutrition security conditions in a given country, multiple indicators have to be used to reflect the various levels and dimensions of the problem.Typical indicators at macro level are: prevalence of common diseases, HIV/AIDS infections, mortality rates and national storage.The presentation of methods to analyse the Food and Nutrition Security situation in a given country is based on this conceptual framework.Targeted food and nutrition security interventions comprise a wide variety of measures which aims at improving food and nutrition security of specific population groups who have been vulnerable or food-insecure.The selection of feasible and (most) suitable approaches of targeted intervention depends on the prevailing conditions, the specific socio-economic features of the target population, the nature and causes of their food deficits and the available reSources and implementation capacities. Pamela et al. (2011) have studied the relation between food and culture.

Targeted interventions have been broadly categorized as

1. Measures aimed at enhancing availability and access entitlement at the individual or household level, like measures to promote food production among small and subsistence farmers, employment and income generation measures, as well as targeted subsidy and transfer systems at the meso level.

2. Measures to improve food utilisation among the target groups, through health and nutrition education, measures of preventive and curative health care, provision of safe potable water, sanitation, promotion of suitable food preservation, food preparation and feeding practices. Although targeted interventions may be defined under macro-level policies, they are usually implemented at meso-level, as projects and programmes in certain areas with a high incidence of food insecure and vulnerable population groups (drought prone or disaster affected areas; low income urban districts; slum and squatter areas). Indispensable prerequisites for targeted interventions are institutional capacities, which can identify and screen the food insecurity for vulnerable population groups, implement the measures to ensure that it reaches out to the eligible target groups effectively and monitor programme performance. The institutions involved may be local self-help groups, community organizations, NGO's, government institutions, international and/ or donor organizations which can have administrative or community approaches. The improvements of productive assets and targeted production support are predominant and are known as the agricultural support mechanisms. It is the small farmers, which constitute the major food insecure population group in many countries. Although respective measures are a part of overall agricultural sector development policies, there is a need to adapt and tailor target the interventions to the specific needs, constraints and capacities of the target group of small and subsistence farmers. Depending on the situation, the following approaches may come under consideration. Water should be safe and wholesome. It should be free from the disease causing agents like bacteria, viruses, parasites and harmful chemical substances like pesticides, industrial wastes, heavy metals, nitrates, arsenic and excess of fluoride. Fluorosis, is a disease which causes bone deformities and dental problems. It is caused due to drinking fluoride rich water over long periods. Generally, a concentration of 0.5 to 0.8 mg of fluoride per litre of drinking water is considered safe. If a water source is not safe for drinking, boiling it for 10-15 minutes is advisable for the purification of water. It kills all disease-causing organisms and removes the temporary hardness. However, boiling will not remove other chemical impurities.

Tablets containing 0.5 g of chlorine can disinfect 20 litres of water. There are many modern gadgets, which claim to provide safe and wholesome water. However, they vary in efficacy. Soft drinks are generally of two categories : natural soft drinks and artificial or synthetic soft drinks. Water is the main constituent of all beverages. Orange, lemon, grape, mango, pineapple and apple are generally used in making fruit juice. Sugar cane juice is also extensively used in India, particularly during summer. In addition to energy natural fruit juices provide some vitamins (beta carotenes, vitamin C) and minerals (potassium and calcium). Fruit juices being rich in potassium are ideal beverages for the hypertensive patients.However, they cannot be equated with fruits which also provide dietary fibre.

According to Yadava et al., (2018a) malnutrition has emerged as one of the most serious health issues worldwide (Yadava et al., 2018 b). The consumption of unbalanced diet amongst poor causes malnutrition. It is more prevalent in the underdeveloped and developing countries. Deficiency of proteins, essential amino acids, vitamins and minerals lead to poor health and increased susceptibility to various diseases which leads to significant loss and affects the socio-economic structure of the country. Although various avenues such as dietary-diversification, food-fortification and medical-supplementation are available, biofortification of crop varieties is considered as the most sustainable and cost-effective approach. In this way, the nutrients reach the target people in natural form. Development of biofortified crop varieties for different nutritional and antinutritional factors is being worked out by the Indian Council of Agricultural Research on the improvement of nutritional traits in different crops. Future prospects of crop biofortification in India are also being discussed. The newly developed biofortified crop varieties are an important source for livelihood and significance in nutritional security to poor people.

Malnutrition poses serious socio-economic implications worldwide, more specifically in the underdeveloped and developing countries (Bouis, 2018). It is usually caused due to inadequate consumption of balanced diet leading to poor health and increased susceptibility to various diseases.In turn it also brings significant losses of annual Gross Domestic Product (GDP), which is as high as 11 per cent in Asia and Africa (Global Nutrition Report., 2017). Globally, around two billion people suffer with malnutrition, while 815 million people are undernourished. Children are most affected due to malnutrition, as a result 151 million children under the age of five were found to be stunted, while 51 million do not weigh enough according to height (wasting) (UNICEF, WHO, The World Bank; 2020). Nearly 45 per cent of deaths among children under the age of five are associated with malnutrition. The problem is so widespread that 88 per cent of the countries experience two or three forms of malnutrition.Among various parts of the globe, Southern Asia is affected the most by malnutrition with 33.3 and 15.3 per cent of the children (<5 yr) are stunted and wasted compared to global average of 22.2 and 7.5 per cent, respectively. In India, 21.9 per cent of the population lives in extreme poverty, and it is estimated that 15.2 per cent of the people are undernourished (Global food policy report, Washington, DC: IFPRI; 2016). As per the National Family Health Survey-4 (2015-2016), 38.4 per cent of the Indian children (<5 yr) are stunted, 21.0 per cent are wasted and 35.7 per cent of the children are underweight. Anaemia is also a serious health issue, where 58.4 per cent of the Indian children (6-59 months) and 53 per cent of the adult women (15-49 yr) are affected from this deficiency.Among adult males 22.7 per cent were found to be anaemic. Considering its widespread ramification, alleviation of malnutrition has been identified as one of the most important steps for hunger-free world. At the beginning of the new millennium, world leaders set eight 'Millennium Development Goals' (MDGs). During 2015, the global community further set 'Sustainable Development Goals (SDGs)' which aim to end malnutrition in all its form (Global Nutrition Report, 2017). Of the 17 goals, SDG-2 (Zero Hunger) aims to end hunger through improved food and nutritional security. SDG-3 (Good Health and Well-being) aims to ensure healthy lives and promotes well-being of people at all ages. Twelve of the 17 goals are highly linked to nutrition thereby signifying its role for better health, education, employment and female empowerment. It has been estimated now that alleviating malnutrition is one of the most cost-effective steps as it offers the benefit of $16 with every $1 invested in any proven nutrition programme (Saltzman et al., 2017).Unbalanced proportion of micronutrients in the diet causes various health problems or designated as 'hidden hunger'.Three major strategies are generally followed to address malnutrition (Gómez-Galera et al., 2010).The best approach would be to increase the diversity of food intake/ dietary-diversification. Among the poor, it is the cereal-based diet which are the mainstay. The inclusion of pulses, fruits and vegetables and even animal proteins in the diet makes the food quite balanced. However, practising dietary-diversification may be difficult in low-income populations. Crop seasonability and low bioavailability of specific micronutrients is another limitation (Gupta et al. 2016). Fortification or supplementation (vitamin A and iron pills/capsules (medical supplementation) or adding nutrients in basic food products like iodized salt (food fortification) have been found very useful. Iron and folate fortified flour and vitamin A added cooking oil are common. According to the Department of Economic and Social Affairs Office for ECOSOC Support and Coordination Achieving Sustainable Development and Promoting Development Cooperation Dialogues at the Economic and Social Council (United Nations, New York, 2008) it has not been found to be sustainable due to the problems in distribution, poor infrastructure and consumer compliance. Biofortification is the most sustainable and cost-effective method for providing the desired levels of nutrients in the diet in natural form (Pfeiffer and McClafferty, 2007). Biofortified crop is a process in which the nutritional quality of a crop is enhanced through genetic manipulation both by breeding and transgenic approaches.

Anti-nutritional factors

Erucic acid: It is a monounsaturated fatty acid found especially in mustard and rapeseed oil. High concentration of erucic acid in edible oils impairs myocardial conductance, causes lipidosis in children and increases the blood cholesterol, <2.0 per cent of erucic acid in oil is desirable for health(Saini et al. 2016; Yadava et al. 2018 a, b).

Glucosinolates: It is a group of thioglucosides mainly found in Brassicaceae family. Glucosinolates produce glucose, sulphate and other aglucon products when broken down by myrosinase. Higher consumption is detrimental to animal health as these reduce the feed palatability and affect the iodine uptake by the thyroid glands, which in turn lowers the feed efficiency and weight gains particularly in non-ruminants like pigs and poultry; <30 ppm of glucosinolates in seed meal is desirable(Yadava et al.2018 a, b; Merah, 2015).

Kunitz trypsin inhibitor (KTI): It is a non-glycosylated protein that possesses adverse effects on growth of humans primarily through inhibition of trypsin in the digestive tract leading to indigestion. KTI constituting the major portion of total trypsin inhibitors in soybean, is considered detrimental to human health (Kumar et al.2018). Although heat treatment is used to get rid of this heat labile protein, the entire amount is not removed.

Biofortified crop cultivars:The National Agricultural Research System (NARS) including Indian Council of Agricultural Research (ICAR) institutes and State Agricultural Universities (SAUs) have contributed immensely to make India self-sufficient in food production. From 50.82 mt of food production in 1950-1951, India has touched 284.8 mt during 2017-2018 (Fourth Advance Estimates; Indiastat, 2018). Similarly, horticulture crops have also experienced increase from 96.56 mt during 1991-1992 to 306.8 mt during 2017-2018 (Third Advance Estimates, Indiastat, 2018). These tremendous gains which have been achieved in yield potential has been possible due to the development of high yielding varieties which were the efforts initiated during Green Revolution. So far NARS has developed 4723 varieties of various field crops. However during the process, nutritional quality was not given importance. Therefore, it has been observed that majority of these varieties do not have good nutritional quality. Looking into the nutritional quality, NARS is developing and releasing many biofortified varieties (All Indian Coordinated Research Projects (AICRPs) for different crops). These biofortified varieties provide calories and essential nutrient(s) which are required for adequate growth and development. Some of the biofortified cultivars released are CR Dhan 310 variety of rice, which is a pure line variety and contains high protein (10.3%) in polished grain. It has been released and notified in 2016 for Odisha, Madhya Pradesh and Uttar Pradesh. Its average grain yield is 45.0 q/ha (quintal/hectare).It matures in 125 days (Yadava et al. 2018 a, b;Yadava et al. 2017).This biofortified variety has been developed by ICAR-National Rice Research Institute, Cuttack, Odisha. DRR Dhan 45, is a pure line variety and possesses high zinc (22.6 ppm) in polished grain. It has been released and notified in 2016 for Karnataka, Tamil Nadu, Andhra Pradesh and Telangana. Its average grain yield is 50.0 q/ha and matures in 125-130 days (Yadava et al., 2018 a, b; Yadava et al., 2017). This biofortified variety has been developed by ICAR-Indian Institute of Rice Research, Hyderabad, Telangana.

DRR Dhan 49, is a pure line variety with high zinc (25.2 ppm) in polished grain. It has been released and notified in 2018 for Gujarat, Maharashtra and Kerala. The average grain yield is 50.0 q/ha. It matures in 125-130 days (Yadava et al., 2018a, b). This biofortified variety has been developed by ICAR-Indian Institute of Rice Research, Hyderabad, Telangana.

Among wheat the biofortified varieties released are ;WB 02, which is a pure line variety and is rich in zinc (42.0 ppm) and iron (40.0 ppm). It has been released and notified in 2017 for Punjab, Haryana, Delhi, Rajasthan (excluding Kota and Udaipur division), western Uttar Pradesh (except Jhansi division), Jammu and Kathua district of Jammu and Kashmir, Paonta Valley and Una district of Himachal Pradesh and Tarai region of Uttarakhand. Its average grain yield is 51.6 q/ha. It matures in 142 days and is suitable for irrigated timely sown conditions (Yadava et al., 2018 a, b;Yadava et al., 2017).This biofortified variety has been developed by ICAR-Indian Institute of Wheat and Barley Research, Karnal, Haryana. HPBW 01, is a pure line variety and contains high iron (40.0 ppm) and zinc (40.6 ppm). It has been released and was notified in 2017 for Punjab, Haryana, Delhi, Rajasthan (excluding Kota and Udaipur division), western Uttar Pradesh (except Jhansi division), Jammu and Kathua district of Jammu and Kashmir, Paonta Valley and Una district of Himachal Pradesh and Tarai region of Uttarakhand. The average grain yield is 51.7 q/ha and matures in 141 days and is suitable for irrigated timely sown conditions (Yadava et al., 2018a, b; Yadava et al., 2017). This biofortified variety has been developed by Punjab Agricultural University, Ludhiana, Punjab, under ICAR-All India Coordinated Research Project on Wheat and Barley. Pusa Tejas (HI 8759) is a pure line variety with high protein (12%), iron (42.1 ppm) and zinc

(42.8 ppm). It is a durum wheat variety suitable for making chapatti (Indian bread), pasta and other traditional food products. It has been released and notified in 2017 for Madhya Pradesh, Chhattisgarh, Gujarat, Rajasthan and Uttar Pradesh. The average yield of this variety is 50.0 q/ha under timely sown irrigated conditions (Yadava et al, 2018 a, b). This biofortified variety has been developed by ICAR-Indian Agricultural Research Institute (IARI), regional station, Indore, Madhya Pradesh. Pusa Ujala (HI 1605) is a pure line variety with high protein (13%), iron (43 ppm) and zinc (35 ppm) and has excellent chapatti making quality. It has been released and was notified in 2017 for Maharashtra, Karnataka and Tamil Nadu. Its average yield is 30.0 q/ha under timely sown, restricted irrigation conditions (Yadava et al, 2018 a, b). This biofortified variety has been developed by ICAR-IARI, regional station, Indore, Madhya Pradesh.MACS 4028 (d) is a pure line durum wheat variety with high protein (14.7%), iron (46.1 ppm) and zinc (40.3 ppm). It has been released and notified in 2018 for Maharashtra and Karnataka. Its average grain yield is 19.3 q/ha under rainfed low fertility, timely sown conditions in Peninsular Zone. It matures in 102 days (Yadava et al.2018a, b).This biofortified variety has been developed by Agharkar Research Institute, Pune, Maharashtra, under ICAR-All India Coordinated Research Project on Wheat and Barley.

Maize : Among biofortified maize varieties; Pusa Vivek QPM9 is an early maturing hybrid which possesses high provitamin-A (8.15 ppm). It also contains high tryptophan (0.74%) and lysine (2.67%) in endosperm protein and is known as (QPM) quality protein maize. It is also a multi-nutrient maize hybrid. It has been released and was notified in 2017 for Jammu and Kashmir, Himachal Pradesh, Uttarakhand (Hills), North Eastern States, Maharashtra, Karnataka, Andhra Pradesh, Telangana and Tamil Nadu. Its average grain yield is 55.9 and 59.2 q/ha, with maturity of 93 and 83 days in the two different agro-climatic zones, respectively(Yadava et al, 2018 a, b; Yadava et al, 2017). This biofortified variety has been developed by ICAR-IARI, New Delhi.

Pusa HM4 improved: It is a medium maturing QPM hybrid that possesses high tryptophan (0.91%) and lysine (3.62%) in endosperm protein. It has been released and notified in 2017 for Punjab, Haryana, Delhi, Uttarakhand (plains) and western Uttar Pradesh. Its average grain yield is 64.2 q/ha. It matures in 87 days (Yadava et al, 2018a, b; Yadava et al, 2017). This biofortified variety has been developed by ICAR-IARI, New Delhi. Pusa HM8 improved variety is a medium maturing QPM hybrid that possesses high tryptophan (1.06%) and lysine (4.18%) in endosperm protein. It has been released and notified in 2017 for Maharashtra, Karnataka, Andhra Pradesh, Telangana and Tamil Nadu. Its average grain yield is 62.6 q/ha. It matures in 95 days(Yadava et al, 2018 a, b; Yadava et al, 2017). It has been developed by ICAR-IARI, New Delhi. Pusa HM9 is a medium maturing QPM hybrid that possesses high tryptophan (0.68%) and lysine (2.97%) in endosperm protein. It has been released and notified in 2017 for Bihar, Jharkhand, Odisha, West Bengal and eastern Uttar Pradesh. Its average grain yield is 52.0 q/ha. It matures in 89 days(Yadava et al, 2018 a, b; Yadava et al, 2017).This biofortified variety has been developed by ICAR-IARI, New Delhi. Besides, Shakti-1 (composite) and hybrids, namely Shaktiman-1, Shaktiman-2, HQPM-1, Shaktiman-3, Shaktiman-4, Shaktiman-5, HQPM-4, HQPM-5, HQPM-7, Vivek QPM-9 and Pratap QPM Hybrid-1 and have also been released for various agroecologies of the country under the AICRP system. These cultivars possess high protein quality due to a higher concentration of lysine and tryptophan.

Among Pearl millet biofortified varieties HHB 299 is a hybrid and possesses high iron (73.0 ppm) and zinc (41.0 ppm) content. It has been released and was notified in 2017 for Haryana, Rajasthan, Gujarat, Punjab, Delhi, Maharashtra and Tamil Nadu. Its average grain yield is 32.7 q/ha, and dry fodder yield is 73.0 q/ha. It matures in 81 days (Yadava et al, 2018 a, b; Yadava et al, 2017). This biofortified variety has been developed by Chaudhary Charan Singh-Haryana Agricultural University, Hisar in collaboration with International Crops Research Institute for the Semi-Arid Tropics (ICRISAT), Patancheru, Hyderabad, Telangana, under ICAR-All India Coordinated Research Project on Pearl millet. AHB 1200 is a hybrid variety and is rich in iron (73.0 ppm). It has been released and was notified in 2017 for Haryana, Rajasthan, Gujarat, Punjab, Delhi, Maharashtra and Tamil Nadu. Its average grain yield is 32.0 q/ha with dry fodder yield of 70.0 q/ha. It matures in 78 days (Yadava et al, 2018 a, b; Yadava et al, 2017).This biofortified variety has been developed by Vasantrao Naik Marathwada Krishi Vidyapeeth, Parbhani, Maharashtra, in collaboration with ICRISAT, Patancheru, Hyderabad, Telangana, under ICAR-All India Coordinated Research Project on Pearl millet.

Among Lentils , Pusa Ageti Masoor is a pure line variety and contains 65.0 ppm of iron. It has been released and notified in 2017 for Uttar Pradesh, Madhya Pradesh and Chhattisgarh. Its average grain yield is 13.0 q/ha. It matures in 100 days and is suitable for the rainfed condition (Yadava et al, 2018 a, b; Yadava et al, 2017). This biofortified variety has been developed by ICAR-IARI, New Delhi. IPL 220, is a pure line variety with high iron (73 ppm) and zinc (51 ppm). It has been released and notified during 2018 for eastern Uttar Pradesh, Bihar, Assam and West Bengal. The average yield of IPL-220 is 13.8 q/ha. It matures in 121 days and is suitable for the rainfed situation during rabi season (Yadava et al.2017). This biofortified variety has been developed by ICAR-Indian Institute of Pulses Research, Kanpur, Uttar Pradesh.

Among the fortified mustard varieties Pusa Mustard 30 is a pure line variety and contains low erucic acid (<2.0%) in oil. It has been released and was notified in 2013 for Uttar Pradesh, Uttarakhand, Madhya Pradesh and Rajasthan. Its average oil content is 37.7 per cent, with a seed yield of 18.2 q/ha. It matures in 137 days and is suitable for timely sown irrigated conditions (Yadava et al.2018a, b; Yadava et al.2017).This biofortified variety has been developed by ICAR-IARI, New Delhi.

Pusa Double Zero Mustard 31; is a pure line variety and contains low erucic acid (<2.0%) in oil and glucosinolates (<30.0 ppm) in seed meal. It has been released and notified in 2016 for Rajasthan (north and western parts), Punjab, Haryana, Delhi, western Uttar Pradesh, plains of Jammu and Kashmir and Himachal Pradesh. Its average oil content is 41.0 per cent, with a seed yield of 23.0 q/ha. It matures in 142 days and is suitable for timely sown irrigated conditions (Yadava et al.2018 a, b;Yadava et al.2017). This biofortified variety has been developed by ICAR-IARI, New Delhi.

Among the soybean varieties NRC-127, is a KTI-free pure line soybean variety. It has been released in 2018 for Madhya Pradesh, Bundelkhand region of Uttar Pradesh, Rajasthan, Gujarat, Marathwada and Vidarbha region of Maharashtra. The average yield of NRC-127 is 18.0 q/ha, and matures in 104 days. It has 18.5-20.0 per cent oil and 38.0-40.0 per cent protein content (Yadava et al.2018 a, b). This biofortified variety has been developed by ICAR-Indian Institute of Soybean Research, Indore, Madhya Pradesh.

Among the Cauliflower fortified varieties Pusa Beta Kesari 1, is a pure line variety and contains high β-carotene (8.0-10.0 ppm). It has been released and was notified in 2015 for the Nation Capital Region of Delhi. Its average yield is 40.0-50.0 t/ha (Yadava et al. 2018 a, b; Yadava et al., 2017). This biofortified variety has been developed by ICAR-IARI, New Delhi.

For Sweet potato many biofortified varieties have been released. Bhu Sona, is a pure line variety and contains high β-carotene (14.0 mg/100 g). It has been released and was notified in 2017 for Odisha. Its average tuber yield is 19.8 t/ha with dry matter of 27.0-29.0 per cent, starch of 20.0 per cent and total sugar of 2.0-2.4 per cent (Yadava et al. 2018 a, b; Yadava et al., 2017). This biofortified variety has been developed by ICAR-Central Tuber Crops Research Institute (CTCRI), Thiruvananthapuram, Kerala.

Bhu Krishna, is a pure line variety and contains high anthocyanin content (90.0 mg/100 g). It has been released and notified in 2017 for Odisha. Its average tuber yield is 18.0 t/ha, with dry matter of 24.0-25.5 per cent, starch of 19.5 per cent and total sugar of 1.9-2.2 per cent (Yadava et al. 2018 a, b; Yadava et al. 2017). This biofortified variety has been developed by ICAR-CTCRI, Thiruvananthapuram, Kerala.

For Pomegranate, Solapur Lal, is a hybrid variety which contains high iron (5.6-6.1 mg/100 g), zinc (0.64-0.69 mg/100 g) and vitamin C (19.4-19.8 mg/100 g) in fresh arils. It has been released and was notified in 2017 for semi-arid regions of the country. Its average fruit yield is 23.0-27.0 t/ha (Yadava et al. 2018 a, b; Yadava et al., 2017). This biofortified variety has been developed by ICAR-National Research Centre on Pomegranate, Pune, Maharashtra.

Globally, the biofortification and biofortified cultivars hold a great promise for health and wellbeing of the human population. Several studies have demonstrated the positive effects of these biofortified crops on humans. A study conducted on 246 children of 12-16 yr of age in Maharashtra, by feeding them with 'bhakri' (round flat unleavened bread) made from iron-rich and conventional pearl millet grains, demonstrated that feeding iron-rich pearl millet was an efficient approach to improve iron status in school-age children (Finkelstein et al., 2015). Serum xanthophylls and retinol were improved to a large extent in Zambia children fed with provitamin-A rich maize grains (Sheftel et al., 2017). In a study with biofortified orange maize it was found as efficacious as a vitamin A supplement in Zambian children even in the presence of high liver reserves of vitamin A which was conducted in a community-based, randomized placebo-controlled trial (Gannon et al., 2014). The beneficial effects of QPM are also well-demonstrated worldwide. Porridge made from QPM resulted in fewer sick days among children compared to those who had porridge from normal maize. Infants and young children fed with QPM expressed 12 per cent higher rate of growth in weight and nine per cent in height compared to the group which was given only normal maize(Gunaratna et al., 2010). Another study showed that consumption of 100 g QPM was sufficient for children to meet the requirement of lysine resulting into reduction in maize to the tune of 40 per cent relative to normal maize (Nuss and Tanumihardjo, 2011). The orange-fleshed sweet potatoes (OFSP) were fed to South African school children aged 5-10 yr, and a favourable response in the vitamin-A status of children was observed compared with the children fed with a traditional white variety of sweet potato (van Jaarsveld et al, 2005). The development and promotion of biofortified varieties thus would be helpful in addressing malnutrition and achieving the SDG's(Paroda and Joshi, 2017). Palmer et al. (2018) have studied the impact of biofortified maize.

Intensive efforts by public sector institutions and policy for intense promotional campaigns can effectively ensure a significant increase in the adoption and acceptance of biofortified crop varieties. Strengthening the seed chain to produce and supply good quality seeds is one of the important steps for the popularization of biofortified varieties. The maintenance of genetic purity is very essential for keeping the quality trait intact.Therefore, it is important that special seed production areas should be identified. Seed village programme can also be one option for taking the seed/commercial production of such varieties to avoid outcrossing from other conventional varieties. Providing subsidized seeds and other inputs would further contribute to the rapid dissemination of nutritionally improved cultivars among the farmers.Assured premium remunerative price through minimum support price for biofortified grains in the market will encourage the farmers to grow more biofortified crops. Investment on extension activities would make the farmers, industry and consumers aware of the availability and benefits of biofortified crops. Many important interventions are required for research, development and for the popularization of biofortified crop varieties. Lack of awareness on the health benefits of biofortified crops is one of the major factors for slow adoption of biofortified varieties. Educational background of the household head and the extent of farmers' participation in demonstration trials and field days are the important factors for the generation of awareness (Gregory and Sewando, 2013; Zuma et al. 2018). The decision by households to adopt the biofortified crops and the subsequent decision to allocate the nutritious food to young children are very significant factors for having a good impact on children's nutrition and health (Zuma et al.2018;Tessema et al.2016). The potential of biofortified varieties give low yields because of which the farmers have apprehension.Therefore, it is not becoming popular among the farmers. However, according to Hossain et al., (2018) and Muthusamy et al., (2014) there are enough evidences to prove that the biofortified varieties are comparable to traditional varieties towards their yield potential. Sometimes, change in color or altered appearance of the produce is happening due to biofortification and this is not allowing the products to become popular. Biofortification of nutrients such as provitamin-A and anthocyanin leads to color change (yellow vs. white maize, and white vs. yellow/purple sweet potato) (Tanumihardjo et al. 2017). Some of the nutritional traits (lysine and tryptophan) are governed by recessive genes and contamination by foreign pollen from neighbouring fields dilutes the quality of the produce in the cross-pollinated crops. This warrants adoption of 'seed village' model for the cultivation of the biofortified crops by the entire farming community of the village. In case of maize, weak linkages between farmers and local poultry companies have been identified as a significant factor for less use of biofortified grains as animal feed (Hellin and Erenstein, 2009). Further, strong linkages with agrifood-processing industry would help in the dissemination of biofortified crops. Strong promotional extension activity such as field demonstration, conveying a message through TV talk, radio shows, flash shows and live-drama would make the farmers, industry and consumers aware of the existence and benefits of biofortified crops.

In order to produce and supply good quality seeds it is important to strengthen the seed chain for the popularization of biofortified varieties of different crops. Providing subsidized seeds and other inputs will be helpful towards the rapid dissemination of nutritionally improved cultivars among the farmers. If the price is fixed through the minimum support price and/or premium price for biofortified grains in the market it will also encourage the farmers to grow more biofortified crops. Recently unveiled National Nutrition Strategy (2017 by NITI Aayog, Government of India) envisages alleviation of malnutrition in the country through food-based solution (Yadava et al.2018a, b). Inclusion of these biofortified cereals in different government sponsored programmes like National Food Security Mission, Rashtriya Krishi Vikas Yojna as well as nutrition intervention programme such as Integrated Child Development Services scheme, 'Mid-day meal' and Nutrition Education and Training through Community Food and Nutrition Extension Units would help in providing the much needed balanced food to poor people. The central government has recently declared millets as 'Nutri Cereals' (sorghum, pearl millet, foxtail millet, finger millet, kodo millet, proso millet, little millet and barnyard millet) besides two pseudo millets (buck-wheat and amaranthus) which have high nutritive value (Yadava et al., 2018a, b). This would boost their demand and allow the farmers to get higher prices. The inclusion of biofortified products in these government-sponsored schemes would especially benefit the children, pregnant women and elderly people and would help in their quick dissemination. Considering the well-demonstrated health benefits of QPM, the Government of Ethiopia has set a target to have QPM varieties cultivated on 20 per cent of the country's total maize area in the coming few years(Tessema et al., 2016).Thus, strong policy support from the government would improve the adoption and acceptance of biofortified crops.

Most of the nutritional traits are phenotypically invisible (protein, lysine, tryptophan, iron, zinc and vitamin C), hence difficult to convince the farmers and traders regarding the extent of quality of the produce (Gupta et al., 2015). Sugar content is measured in the harvest of sugarcane and sweet corn with a brix meter. The development of easy to use equipment to give reliable estimates of the quality parameters of the produce would be beneficial. It is also vital that research work undertaken should be directed towards newer traits which are helpful for nutrition security. Towards this, ICAR has recently funded a Consortia Research Platform on 'Biofortification in selected Crops for Nutritional Security', for the development of biofortified crops. The work is underway in which rice, wheat, maize, pearl millet, sorghum and minor millets have been targeted for nutritional enrichment (Neeraja et al., 2017). Phytic acid has been identified as one of the most important anti-nutritional factors that reduces the bioavailability of iron and zinc in our body. Andersson et al.,

(2017) have explained that in the human gut, only 7-7.5 per cent of iron and 15-25 per cent of zinc in food grains are available for absorption. On the activity of phytase enzyme a lot of genetic variation has been found which is responsible for the degradation of phytic acid in wheat and specific germplasm lines with high phytase activity. This variety is being used as donors in the breeding programme. In rice and maize, low-phytic acid mutants (lpa) that lead to lower synthesis of phytic acid are being used in the breeding progarmme (Neeraja et al. 2017). Similarly, high oil trait that enhances bioavailability of provitamin A in maize is under study.There is also a need for breeding varieties which would enhance ascorbic acid and β-carotene which will indirectly increase the bioavailability of iron and zinc (Welch and Graham, 2004). Addressing the nutrient deficiency and having solutions by crops seems to be a viable option. Multinutrient rich cultivars like 'Pusa Tejas' a wheat variety which is rich in protein, iron and zinc, 'Pusa Vivek QPM9 Improved' variety of maize which is rich in provitamin A, lysine and tryptophan and Solapur Lal which is a pomegranate variety containing high vitamin C, iron and zinc, have been developed. Maize genotypes which have high lysine, tryptophan, provitamin A and vitamin E have been developed and are under evaluation at many different locations. The utilization of modern tools and techniques should be the integral part of all breeding programmes. Genomics-assisted breeding coupled with doubled haploid and speed-breeding technologies will accelerate the development of biofortified cultivars to a large extent in the coming times. Wherever sufficient genetic variation for the target trait is not present in the germplasm, then transgenic approach plays a significant role. Using the technology, NARS, has developed 'golden rice' with high provitamin A. There needs to be a better coordination and collaboration among the breeders, biotechnologists, biochemists, seed technologists, agronomists and post-harvest technologists across various public and private organizations which will help in the development of nutritionally rich crop varieties in an effective manner.Priority should be given for efficacy trials on the Indian population to envisage the effects of biofortified varieties. Under the ICAR-CRP on Biofortification, ICAR-IARI, New Delhi in collaboration with ICMR-National Institute of Nutrition, Hyderabad have analyzed the bioaccessibility of provitamin A maize using *in vitro* model system and found the results to be very encouraging(Neeraja et al. 2017). Scientists have taken up work on the human clinical trials that will evaluate the effects of provitamin A rich maize on humans. Provitamin A rich maize fed chickens will be an effective way to improve vitamin A consumption among non-vegetarians populations through eggs (Liu et al.2012). The ICAR-IARI, New Delhi, in collaboration of ICAR-Directorate of Poultry Research, Hyderabad have initiated a programme to evaluate the effects of provitamin A rich maize on indigenous poultry breed,

Vanaraja. Large-scale trials will contribute in dissemination. The high yielding biofortified crops will contribute significantly towards nutritional security. Prof. M.S. Swaminathan father of Green Revolution, has rightly said that we need to move from food security to nutrition security, where not only calories and proteins but also micronutrients are taken care. There is an urdent need to bring agriculture, health and nutrition together.

It was Jackson et al., (2018) who made a study in Sub-Sahara African area, Congo , the area, which suffers with both malnutrition and food insecurity. The most vulnerable groups include the young children, lactating and pregnant mothers. The use of the starchy staple foods like cereal-based foods and cassava are very common. These have low protein content and are deficient in micronutrients. Therefore, if traditional and nutritious foods like mushrooms can be enriched with high protein, micronutrients and fiber it will be helpful in improving the health of the people. Mushrooms are ranked to be richer than most food sources except meat in term of protein content (Jackson et al.2018). They are high in nutrition and are low in cholesterol content. This work will help to combat the protein malnutrition and micronutrient deficiencies as a novel food fortification approach (Ishara et al. 2018). Recently, food fortification has gained a lot of momentum in India to address the prevalence of vitamin and mineral deficiencies among India's 1.3 billion population. The Haryana state government began distributing fortified atta in two blocks of one district, reaching approximately 177, 000 people with iron, folic acid and vitamin B-12. Haryana's government plans to scale it up so that it could reach to 12 million people across the entire state.The work is in line with World Health Organization recommendations for wheat flour fortification along with the FSSAI standards (Haryana project, 2018;Annual Report).Fortifying wheat flour through appropriate market channels could be replicated in 17 states in India so as to reach 400 million people who are highly vulnerable to vitamin and mineral deficiencies. Atta is whole-wheat flour used to make chapaties and rotis.Fortifying these commonly consumed foods with folic acid may reduce the number of babies with brain and spine birth defects and the number of children and women with nutritional anemia.According to the Food Safety Standards Authority of India (FSSAI, 2016), food fortification is a realistic and sustainable complementary strategy to food supplementation and dietary diversification to eliminate micronutrient deficiencies. Foods for fortification, according to the declaration by FSSAI are mainly; milk, edible oil, rice, salt and wheat flour.

To have the greatest health impact, grain fortification in India must consider both wheat flour and rice. The amount of rice and wheat in India's food supply is almost equivalent, according to the Food and Agriculture Organization

(FAO) of the United Nations. In 2013, the most recent year for which data are available, the amount of milled rice in the national food supply was 190 grams per person per day; the amount of wheat was 166 grams per person per day. Fortified rice is available through social safety net programs in the states of Odisha and Karnataka. Also, fortified rice is being distributed in two research trials in India which include Tamil Nadu and Gujarat. Anemia is prevalent in twenty-four states of India.It ranges between 26 to 65% among married women, however, the average is 50%. Prevalence over 40% comes under severe public health concern by the World Health Organization. It can lead to fatigue, limits cognitive development in children and contributes to maternal deaths.In India, out of every 10, 000 births around 45 have a defect either of the brain or of the spine. Out of 25.6 million annual births, 115, 390 numbers are found with birth defects of the brain or spine. It has been estimated that if there is an adequate intake of folic acid (a form of vitamin B9) then it could prevent more than 100, 000 deaths among children under five years of age every year. These birth defects are called as neural tube defects. The '+F' logo has been notified to identify fortified foods. Food Safety and Standards of India FSSAI, under the Fortification of Foods Regulations, (2018) have notified in the Gazette of India on 09.08.2018 that fortification is the addition of key vitamins and minerals like iron, iodine, zinc, vitamins A and D to the staple foods like rice, wheat, oil, milk and salt to improve their nutritional content. These nutrients may or may not have been originally present in the food before processing or may have been lost during processing. As India has a very high burden of micronutrient deficiencies caused by vitamin A, iodine, iron and folic acid it is leading to diseases like night blindness, goitre, anaemia and many birth defects. According to the National Family Health Survey (NFHS-4) 58.4 percent of children (6-59 months) are anaemic, 53.1 percent women in the reproductive age group are anaemic, 35.7 percent of children under 5 are underweight. Fortification is a globally proven intervention to address the much prevalent micronutrient deficiencies in the population.Food Fortification has a high benefit-to-cost ratio. The Copenhagen Consensus has estimated that every one rupee spent on fortification results in nine rupees to the economy benefit. It requires an initial investment to purchase both the equipment and the vitamin and mineral premix, but the overall costs of fortification are extremely low. Even when all program costs are passed on to consumers, the price increase is approximately by 1-2%, which is less than the normal price variation. The benefits of fortification of foods are that the nutrients are added to the staple foods since they are widely consumed and it becomes an excellent method to improve the health of a large section of the population. It is a safe method of improving nutrition among people. The addition of micronutrients to food does not pose a health risk to people and the quantity added is small that falls under the Recommended Daily Allowances (RDA) and are well regulated as per prescribed standards for safe consumption. It is a cost-effective intervention and does not require any changes in eating patterns or food habits of people. It is a socio-culturally acceptable way to deliver nutrients to the people. It does not alter the characteristics of the food like the taste, aroma or the texture of the food. According, to FSSAI, the food business operators (FBOs) need to comply with the provisions of these regulations within a year. The new standard provides a minimum and maximum range for fortification of staples like wheat flour (atta), maida, rice, salt, vegetable oil and milk. The dosage of the micronutrients are adjusted so as to provide 30 to 50 per cent of the daily requirements. In milk and oil, the unit of dosage has been changed to microgram retinol equivalent for vitamin A and microgram for vitamin D from IU. In wheat flour and rice fortification, other sources of iron have been added, while vanaspati fortification has been excluded. According to the country's apex food regulator, the adoption of the standards by 62 top companies and their brands of all five fortified staples are available in the open market across the nation. Leading the movement, the oil and milk industry, with 47 per cent packaged refined edible oil industry and 21 per cent of the organised milk industry is fortifying their products as per FSSAI standards. As per FSSAI and Food Fortification Resource Centre (FFRC), fortified staples (wheat flour, oil and DFS) are being used in Integrated Child Development Services (ICDS) and the Mid-Day Meal (MDM) scheme. Ministry of Women and Child Development and Ministry of Human ReSource Development have made it mandatory. Presently, Odisha, Karnataka, Haryana, Gujarat, Uttar Pradesh, Himachal Pradesh, Madhya Pradesh, Maharashtra, Jharkhand, Rajasthan, Tamil Nadu, Tripura, West Bengal, Kerala and Andhra Pradesh and three Union Territories (UT's) have adopted fortification of their chosen commodities at the district or at scale in the government safety net programmes (SNP), namely ICDS, MDM and the public distribution system (PDS). FSSAI set up FFRC with the support of TATA Trust to promote and support large scale implementation of food fortification in the SNP as well as ensure availability of fortified products in the open market. FFRC provides end-to-end technical support as well as advocacy in collaboration with development partners, like TATA Trust, GAIN, PATH, the World Food Programme (WFP), World Bank and NI. Companies in the packaging business are likely to partner with small- and mid-tier food producers to offer solutions for incorporating the new packaging and labelling requirements.

11

Recent Innovations in Instant Mixes of Traditional Foods for Nutritional Security

A number of innovations have taken place towards ready to eat instant food mixes of traditional origin. These have helped people to eat traditional food in which ever part of the world they go.

MTR Company

In India the best example is of MTR company which manufactures a range of packaged foods which includes breakfast mixes, ready to eat meals, masalas and spices, snacks and beverages. Snacks, such as kodubale, khara boondi and masala groundnut are very popular as they are ready to eat packets and contain healthy food with less oil. The different spices used for mixes are procured from famous countries which give the perfect taste and avoid rancidity.As a summer launch, milkshake with canned chikoo, banana and cold filter coffee are available. MTR Foods Pvt. Ltd. is a subsidiary of Norwegian conglomerate Orkla and is an acronym of Mavalli Tiffin Rooms. According to the company, India is on the verge of explosive growth in packaged instant foods. MTR has an aim to target a market of Rs 30, 000 crore with their products of traditional meals, breakfast, snacks and desserts. The Nielsen India study has shown that in 2020, the number of consumers in this category will be 2.7 times than the current base.

Mother's Recipe is a brand which make traditional pickles in bottles with a distribution network to 20 states in India and exporting the products globally. The manufacturing is done in-house and the raw materials are procured seasonally in large quantities. After cleaning they are cut according to the recipe. Each variety of mango pickle has its own cutting process. Some are automatically cut, while others need manual work. The use of right oils, spices roasted inhouse, and marination of the pickles according to the practices of the particular region allow many consumers to relish them. Bottling and packaging are done only when the orders come.

Britannia Industries Limited: Britannia company was founded in 1892 and is the largest producer and seller of biscuit products in India. This company is based in Kolkata, West Bengal and is known as 'The King of Biscuits'. Britannia company delivers a wide range of food products like Good Day, Britannia Cracker, Britannia nutria Choice, Britannia Little Hearts, Britannia bread and Britannia fruit-cakes, which are fresh and full of nutrients.

Parle Products: The famous Parle-G, Monaco, KrackJack biscuits and many other confectionery such as Poppins, Melody, Mango Bites, are produced by Parle Products. It is one of the largest Indian food companies. The company was founded in 1929 and has been catering biscuits to Indian homes since 1939. Their ad campaign in the newly independent India in 1947 showcased Gluco biscuits as a desi alternative to the biscuits made by the British.

Hatsun Agro Products Ltd.: The company which was started in 1970, has been a pioneer in promoting dairy products. It is the largest private sector dairy in India. Dairy ingredients are exported to 38 countries. It holds various quality certifications. This company has tied up with banks and has arranged animal loans for dairy farmers. It has established an extremely efficient supply chain management. The skimmed milk powder is very light, delicious and healthy for consumption. It is made from the purest of the milk collected from the cows and is 99% fat free. The products are fat free, full of milk goodness and easy to digest. The skimmed milk powder is rich in protein, cholesterol free with no added sugar content.

Kwality Wall's: Kwality Wall's is famous for its delicious ice-creams. Having founded in 1956, the Kwality Wall's ice-cream is now a household name in India. It's headquarter is in Mumbai and its popular products are cornetto, feast, creamy delights and cassatta.The cheese cake contains real strawberry pieces and a decadent strawberry sauce. The icecream is delightful in taste and smooth in texture. It is creamy and gives the effect of real strawberry chunks with each bite.

Rasna: India's one of the leading manufacturers and exporters of beverages is Rasna, which has an effective nutritional value, superior natural taste and a maximum number of flavors. This company is based in Ahmedabad, Gujarat. Its main products are instant drink, frutants, syrups, fruit jam and fruit drink. Aam Panna is an instant drink concentrate mix. The drink mix is free of any unwanted chemicals and is all natural. The Aam Panna can be quickly made in 5 minutes. The goodness of mango helps in energising and maintains the required water level in the body. The concentrated mix helps in the flawless functioning of the immune system. Rasna has perfected their blend with 21 vitamins and glucose.

Rei Agro: Rei Agro was established in 1996. It has consolidated the fragmented basmati industry. Its headquarter rests in Kolkata, West Bengal. It is the largest producer of processed basmati rice in the world. Raindrops basmati rice comes from the house of rei agro Ltd.

Kissan : The company was established in 1935 in India. Kissan has become the first ever fruit and vegetable brand in India. This company introduced new formats of food such as canned fruits and vegetables and baked beans. Innovation is the main approach in the case of kissan.One jar of mixed fruit jam is full of all favourite fruits. One scoop provides good nutrients. It is a medley of eight fruits that includes; banana, papaya, apple, pear, pineapple, mango, grape and orange which are mixed together in to one delicious jam. The jam can be used in multiple ways.

LT Foods: LT foods company was launched in 1980. Its flagship brand 'Dawat' is now recognized as the leading brand in the ready-to-eat food industry. The products of this company are branded rice, wheat and pulses, healthy snacks, spices and other value-added products like brown flex seeds and cashew nuts. In India, this company has a strong nationwide distribution network.

Modern dairies: The company is in Karnal, near Delhi. This company uses ultra- modern machinery, supplied by reputed manufacturers. Modern dairies cover a variety of milk and milk products for the domestic as well as international markets. The malai paneer of the brand is silky in texture and taste. The white block is full of protein, fibre and calcium. The paneer is made out of fresh and rich milk that results in nutrient cottage cheese.

Heritage Foods Ltd.: It is one of the leading food processing companies in India. The heritage group was established in 1992.The fastest growing private sector enterprises in India, the heritage group, has five business divisions, namely, dairy, retail, agro bakery and renewable energy. Heritage Badam milk is a milk based drink which is full of nutrients and is balanced with required amount of carbohydrates and proteins. It also has the suggested amounts of bone boosting calcium, phosphorous and vitamin D.

Kohinoor Foods Limited: It is a global Indian food processing company. The company was established in 1989. The company offers one of the finest basmati rice brands. Other food products of this company include wheat flour, ready- to- eat curries and meals, simmer sauces, cooking pastes to spices and frozen food.

Ruchi: Ruchi Soya is the country's largest solvent extraction and edible oil refining company.It has emerged as the leader of India's edible oil industries. It has 5 port- based refineries, three standalone crushing plants, eight integrated crushing and refining plants, one refinery and vanaspati plant and two palm fruit processing units. The company's brands are nutrela, vanaspati and sunrich.The soya chunks are made up of 100% flour of soy beans. Soy bean products are claimed to be an effective substitute products for people who do not want to derive their protein from meat and eggs. Soy chunks are 100% pure vegetarian item full of protein and other vitamins. They can be used in any gravy to alleviate the taste and make it more nutritious. As compared to meat and eggs the protein content in soy chunks is 54%. It is advisable for a healthier heart because of low fat and cholesterol free traits.

Amul: The company was founded in 1946. It is managed by the Gujarat co-operative milk marketing federation ltd. Amul, the co-operative farm, was formed as a response to the exploitation of marginal milk producers by traders. The major products of Amul company are milk, butter and cheese. Being the largest milk producer in the world, amul is marketed by 22 state marketing federations. Butter is a very famous product.

Dabur Foods Ltd.: Dabur India Limited is the fourth largest FMCG company in India. For the past 125 years, the Dabur company has been providing nature based solutions for a healthy and holistic lifestyle. The food products include fruit juices, spice pastes, sauces, refreshment drinks and lemoneez. Dabur honey, is an excellent replacement for refined sugar with many nutrients.

Godrej beverage and foods Ltd: A joint venture between the Hershey company and Godrej group named GHL was formed in 2007. Products in confectionery, beverages and grocery items are famous. Nutrine confectionery company is a key portfolio for GHL. The beverage portfolio includes fruit drinks, juices and soya milk. Yummiez chicken nuggets are fried but lightly breaded. The nuggets are juicy, tender and crispy when fried. The yummiez nuggets are made up of fresh meat. The ingredients used are easy and completely free of all types of chemical preservatives or fillers.

Lakshmi Energy: It is a food processing company in India and is based in Punjab. Wheat and rice are the main products. This company grows 40% of rice and 70% of wheat, grown in India. Laksmi energy delivers products like basmati and non-basmati rice, rice bran oil and wheat flour. The wheat crops are selected after multiple screenings.

Vadilal Industries Limited: The Vadilal Company is an Indian food processing company. It serves its customers with the high quality and best products. The products of this company are ice-cream and processed food. The Vadilal Golden Fantasy Cake is a ice cream cake which has a nutritive toppings of dry fruits.The products are made up of pure cow milk making it healthier.

Kellogg's: From cereals to snacks, Kellogg's is one of India's oldest instant food brands that started with cornflakes in 1994. Now they make a variety of healthy and tasty breakfast cereals with oats and muesli. They also have a range of instant snacks.

McCain: Fairly new to the Indian market, McCain foods is a subsidiary of its Canadian counterpart and has been involved in agricultural R&D since 1998 in India. Today, McCain provides a range of delicious ready-to-eat frozen food like McCain French fries, Aloo Tikki, Smilies, Cheesy bites, and other delectable snack items that needs to be unfreezed and fried.

TC India (Kitchens of India): With brands such as Kitchens of India that sell ready-to-cook curries and rice dishes.

Nestle (Maggi): Maggi Noodles has been a staple childhood diet. It is a quick, easy and India's most popular ready-to-eat processed foods. Many varieties available for everyday use include wheat noodles, cup noodles, soups, soup cubes, pasta and instant masala mixes.

12

Future of Traditional Foods and Nutritional Security Through SMEs

Benjamin et al., (2016) have focused on the small-scale food processing enterprises (SSFPEs) as one of the important measures for national development and addressing food security challenges, particularly in Nigeria. The authors have explained the need for food processing as there has been tremendous food losses and wastages in the country. Enhancing the food processing and productivity for mechanization of food processing unit operations, development of new and existing technologies, design and development of machinery, systems for processing and preservation of different agricultural produce of high target have been undertaken. SSFPEs have played an important role in the economy of a developing country, particularly in terms of employment creation, income generation, post-harvest losses reduction, food preservation, value addition, improvement of food safety, nutritional quality, increase in shelf-life of a product and to act as training grounds for entrepreneurs before they invest in large scale enterprises. SSFPE's around the world and especially in Nigeria are faced with significant challenges that compromise their ability to function and contribute optimally to the economy. Financial constraints and lack of management skills are the major constraints. It is important that governments and non-governmental organizations must encourage and facilitate the development of SSFPE's. The mention of specific companies or products of manufacturers, whether or not these have been patented, does not imply that these have been endorsed or recommended by FAO or GAIN in preference to others of a similar nature that are not mentioned. FAO's Nutrition and Food Systems Division (ESN), in collaboration with the Global Alliance for Improved Nutrition (GAIN) 2018, conducted a workshop on Leveraging Small and Medium Enterprises (SMEs) to improve nutrition (FAO, 2018; ISBN 978-92-5-131212-4). The workshop was organized with the support of project (strengthening capacities for nutrition sensitive food systems through a multistakeholder approach funded by the Ministry of Agriculture, Forestry and Fisheries (MAFF) of Japan and the FAO Multipartner Programme Support Mechanism (FMM) 2015–2017). This enables the women to benefit more from the agri-food value chains. Making nutritious and safe foods available, accessible, affordable and desirable for improved nutrition requires action from all stakeholders in the food system, including producers, processors, retailers and consumers both in private and public institutions. SMEs in the food industry play a very major role, as they provide a large share of the day-to-day sustenance for low-income households. But they are overlooked, due to the policies on agriculture, food safety, nutrition, trade and agro-industry. It is important to know if the market can produce and sell foods that contribute towards better health. Can 'nutrition' be a business opportunity for SMEs. Can it be done in a sustainable manner. To work towards higher demand for these foods by using the markets for facilitating the supply of nutritious foods by reducing the production costs. SMEs struggle to make nutritious foods which are affordable to poor consumers in an economical manner. In order to reduce the costs the tax system should be more supportive of SMEs that produce nutritious foods (reduce import tax on ingredients for micronutrient fortificants), enable an affordable, adequate and steady supply of quality ingredients through agricultural policies that support the production of more diverse and nutrient-rich foods, extend agricultural subsidies beyond the farm to support the food system. SMEs in particular processes nutritious foods, stimulates technology and innovation that removes the obstacles in the supply chain and are adapted to the scale of SMEs. They provide training and streamline food safety standards and regulations and facilitate the transfer of technologies, facilitate access to necessary inputs such as credit, infrastructure, energy and reduce administrative barriers for SMEs and facilitate essential business procedures. Another important aspect is to look into the demand for nutritious foods and marketing as it is important for the sustainability of SMEs, as they lack the marketing reSources as compared to the larger companies. Recently, suggestions have been made to conduct public campaigns for nutritious foods, inform consumers of their benefits, increase access of SMEs to market analyses and consumer research, regulate labelling claims and marketing strategies, to preclude abusive nutritional claims and unfair competition and improve the links between SMEs that produce nutritious foods and public procurement mechanisms. SMEs are too big to micro-finance, but, at the same time, they are small for commercial lending. Therefore, it is important to bring in newer ways which can help towards access to finance at the time of necessity by any entrepreneur.Change is required in credit schemes so that they could be accessible and adapted to SMEs (including with government support, in the context of PPPs). Work should be done in a direction to explore the possibility of a 'Nutrition Fund' that will support the SMEs towards nutritious foods. These should be integrated with large public investment funds like the Global Agriculture and Food Security Program (GAFSP) and national investment plans (National Agriculture Investment Plans, NAIPs; Herforth and Ahmed

2015). Many of the SMEs which supply nutritious foods are socially-oriented companies that contribute towards the better nutrition of the population. SME representatives wanted to have easier policies and programmes so that their businesses becomes viable. In a Small and Medium Enterprise fewer number of employees are taken depending on the country. European Union has fixed as 250 employees. It is also important to reach the low-income consumers and create a market for SMEs producing nutritious foods through public institutional procurement systems, like the school feeding and food assistance programmes. The strength of the SMEs will be if they can be brought closer to low-income consumers. These actions could be facilitated, if, government would place nutrition on the top in the political agenda and in food and agriculture.It must be brought to the mainstream investment. The policy makers have a pivotal role to play on the policies and procedures to create an enabling environment. Good policies must be shared across the countries and among institutions. Research and development needs support especially for the use of alternative energy reSources, research on underutilised nutritious foods and for developing a curricula for the next generation who are working in the food sector. Public education and communication campaigns need to be organized on nutrition and healthy diets. Regulation of labelling and marketing must be done in order to prevent abuse and the spread of misinformation. It is also very vital to follow the rules made under codex Alimentarius. Monitoring of food consumption patterns among people will bring understanding and more attention. Advertisement, generation for a demand of nutritious foods and regulations for labelling and marketing will be worthwhile. It is also worthwhile to encourage the greater use of local nutritious foods which will help the poor consumers. A good network should be made in terms of information and data.Nutritional issues must be addressed via agriculture and rural development investment plans and that interventions are implemented in collaboration with SMEs producing nutritious foods.Opportunities must be explored for partnerships with local commercial banks and international financial institutions which can promote SMEs. SMEs advocating with governments, commercial banks and international financial institutions must develop a method to lend products adapted for their needs. Nutrition should be integrated with large agriculture and rural development plans which have partnerships with the local SMEs. It will be a step forward for the SMEs to associate with Finance and Investment Network (SAFIN) which are supported by IFAD and FAO (De la Peña et al. 2018).

13

Conclusions

A sustainable and healthy food system is one which can provide food and nutrition security for all, in an economical way which is socially and environmentally sound for future generations. Underutilized indigenous and traditional crops can support and strengthen the existing food system, as they are in particular economical and friendly with the environment. Several underutilized indigenous and traditional crops are nutrient dense and adapted to marginal conditions, suggesting that they could be used to champion sustainable and resilient agriculture and food systems for smallholder farmers residing in these environments. The mainstreaming of underutilized indigenous and traditional crops and foods should be recognized for their attributes to make them desirable and to conserve them. In this regard, the fact that value chains for underutilized indigenous and traditional crops are poorly developed at present has created many opportunities to diversify the current food systems. This will create a new employment, market and distribution opportunities.It will also help to combat poverty. It will be if the smallholder farmers and women in rural areas are given the opportunity to utilize underutilized indigenous and traditional crops and re-enter the manistream system. A significant role which is played by the women for the production and conservation of underutilized indigenous and traditional crops offers opportunities for women empowerment through their inclusion in the food system. Promotion of gender equality and empowerment of women is directly linked for strengthening the sustainable food systems that can fight hunger and malnutrition and improve the lives and livelihoods of rural populations. The mainstreaming of underutilized indigenous and traditional crops into the food system would support women to diversify in a sustainable way, feed their own households and to provide nutritious food into the local markets. This will also help in addressing, the Sustainable Development Goals which are related to social, economic and environmental issues. Specifically, SDG's 1, 2, 3, 8 and 15. Due to the low literacy and numeracy skills of the smallholding farmers it hampers an efficient production and also the adoption of newer technologies. Therefore, it needs a strong training component to address their underdeveloped skills. Initiatives must be taken to integrate smallholder farmers into the agro-industrial food systems. There is a positive correlation between farmer training and technology adoption. Most of the times, it results in positive economic and environmental outcomes. The researchers and the policy makers should pay attention to these sectors with a nexus approach, and not to view them separately. In the past there have been so many inequalities and power imbalances, especially apartheid discriminative policies which has brought down the diversity of crops, reduced access to the food system and reduced the diversity of diet. All these have ultimately reduced the household food security and has increased the hidden hunger, increasing the malnutrition and consequences associated with it. When we include underutilized indigenous and traditional crops into the food system, it increases the diversity in crops, improves the local value chain and diversifies food. This ultimately, increases the household food security, improves livelihood and reduces hidden hunger. The current policy environment makes little mention of underutilized indigenous and traditional crops and is unfavorable for growth and process in informal food systems. In addition, the current policy environment leaves critical aspects of the food system like marketing and distribution to the private sector which may not be promoting traditional food. It leaves the holders of the indigenous and local knowledge about these diverse crops vulnerable to corporate patents that do not take into account access and benefit sharing legislation. It is important to include the scientists (at the start of the policy value chain), and the beneficiaries (at the end of the value chain). Indigenous knowledge systems are necessary as they provide an alternative information source to western-driven science. It needs to be made accessible and translated for decision-makers. The different power relations in the food system must be recognized, as these may influence the success of any interventions. As envisaged in the Sustainable Development Goals for better health, wealth and education which will have long-term effects on national development and nation building it is important to take traditional food systems in to the main stream. Future studies should focus on conducting an econometric assessment to determine a cost-benefit analysis for mainstreaming underutilized indigenous and traditional crops. Studies in Africa are available online at http://www.mdpi.com/2071-1050/11/1/172/s1.

References

Abil TA, Elegalam A. 2007. Investigation into the chemical composition of the dry fruit of Tetrapleura tetraptera (Ubukirinu). J Fd Technol. 5:229–232.

Aboaba OO, Ezeh A R, Anabuike C L. 2011. Antimicrobial activities of some Nigerian spices on some pathogens. Agri Biol J North Amer. 2: 1187–1193.

About Food in Malaysia. 2012. Travellers Worldwide. Archived from the original on 21 June, 2012. Retrieved 1 July 2012.

Abukutsa MOO. 2007. The Diversity of Cultivated African Leafy Vegetables in Three Communities in Western Kenya. Afri J Fd Agri Nutri Develop. 7: 3. ISBN 1684-5374.

Achaya K T. 2003. The Story of Our Food. Universities Press. Retrieved 18 June 2015.

Achi O K. 2006. Composition and antibacterial activities of Tetrapleura tetraptera Taub. Pod extracts. Res J Microbiol. 1:416–422.

Acquah B, Kapunda S and Legwegoh A. 2014. The dimensions of urban food insecurity in Gaborone, Botswana. Urban Forum. 25: 217–226.

Adam, C and Dercon S. 2009. The political economy of development:An assessment. Oxford Rev Econ Pol. 25. 173-189. 10. 1093/oxrep/grp020.

Adamkowska H. 2008. Kuchnia. Podróże kulinarne. Polska. Tradycje, smaki, potrawy. Rzeczpospolita. (Kitchen. Culinary travels. Poland. Traditions, flavours, dishes), New Media Concept.

Addo AA and Ogumntona CRB. 1993. Nutritional value of soybean. A paper presented at Training Workshop of Extention Workers in Soybean Processing and Utilization. FMAWAR and /RS/UNAAB. 9.

Adebooye O C and Opabote J T. 2004. Status of conservation of the indigenous leaf vegetables and fruits of Africa. Afri J Biotechnol. 3: 700-5.

Adedipe NO, Odegbaro OA, Aliyu A and Jir M. 1995. Integrating agricultural training and research in Nigeria. Nat Agric Res Proj. Monograph No 2: 169

Adefegha S A and Oboh G. 2012. Acetylcholinesterase (ache) inhibitory activity, antioxidant properties and phenolic composition of two aframomum species. J Basic Clin Physiol Pharmacol. 23:153–161.

Adefegha SA, OlasehindeTA and Oboh G. 2017. Essential oil composition, antioxidant, antidiabetic and antihypertensive properties of two Afromomum species. J Oleo Sci. 66(1):51-63. doi: 10. 5650/jos. ess16029. Epub 2016 Dec 8.

Adesina S K, Iwalewa E O and Imoh I J. 2016. Tetrapleura tetraptera Taub- ethnopharmacology, chemistry, medicinal and nutritional values-a review. Brit J Pharma Res. 12:1–22.

Adetunji J A. 2007. Tetrapleura tetraptera: Molluscidal activity and chemical constituent. AfriJ Trad Compl Altern Med. 4:23–26.

Adeyemi M M H. 2011. A review of secondary metabolites from plant materials for post harvesting. Intern J Pure Appl Sci Technol. 6:94–102.

Adrian D. 2017. An overview of agriculture, nutrition and fortification, supplementation and biofortification:Golden rice as an example for enhancing micronutrient intake. Agri Fd Sec. 6:59.

Agence Française de Sécurité Sanitaire des Aliments (AFSSA). 2010. Agence française de sécurité sanitaire des aliments relatif à l'actualisation des apports nutritionnels conseillés pour les acides gras Saisinen 2006-SA-0359. Maisons-Alfort, France.

Ahmad N H, Noorul K, Maad A, Juber M, Mohd B M, Anuradha K, Nazma. 2016. Glycyrrhiza glabra: for traditional uses and pharmacological actIONS Corresponding author. Adv J Pharm Life Sci Res (AJPLR). 2. 23-32.

AICR, 2007. Joanna Kruk. Health and Economic Costs of Physical Inactivity. DOI:http://dx. doi. org/10. 7314/APJCP. 2014. 15. 18. 7499.

Alaje DO, Owolabi KT, Olakunle TP, Oluoti O J, Adetuberu IA. 2014. Nutritional, minerals and phytochemicals composition of Garcinia cola [Bitter cola] and Aframomum melegueta [Alligator pepper]. IOSR J Environ Sci Tox Fd Technol (IOSR-JESTFT). 8:86–91.

Alan D. 2014. Tom Jaine (ed.). The Oxford companion to food (3rd ed.). Oxford: Oxford Univer Press. 410–411. ISBN 0-19-967733-6.

Alderman, Hoddinott and Kinsey. 2006. Long term consequences of early childhood malnutrition. Oxford Economic Papers. 58(3):450–474. https://doi. org/10. 1093/oep/gpl008

Ali BH, et al. 2008. Some phytochemical, pharmacological and toxicological properties of ginger (Zingiber officinale Roscoe): A review of recent research. Fd Chem Toxic.

AlkireS , Meinzen-DickR, PetermanA, QuisumbingA , Seymour G and VazA. 2013. Women's empowerment in agriculture index. World Dev. 52: 71-91.

Amazingarunachal. com. Retrieved 4 June 2012. Arunachal Pradesh staple food.

American Dietetic Association. 2007. Healthy land, healthy people. Building a better understanding of sustainable food systems for food and nutrition professionals. A primer on sustainable food systems and emerging roles for food and nutrition professionals. Chicago, IL: American Dietetic Association. BY:Angie Tagtow et al.

Amugsi D , Lartey A, Kimani E and Mberu B. 2016. Women's participation in household decision making and higher dietary diversity: findings from nationally representative data from Ghana. J Health Popul Nutr. 35 (1):16. https://doi. org/10. 1186/s41043-016-0053-1.

Anand K. 2012. Treats from the Valley. The Times of India.

Anand SS, Hawkes C, De Souza RJ et al. 2015. Food consumption and its impact on cardiovascular disease: importance of solutions focused on the globalized food system: a report from the workshop convened by the World Heart Federation. J Am Coll Cardiol 66:1590–1614.

Andersson MS, Saltzman A, Virk PS, Pfeiffer WH. 2017. Progress update: crop development of biofortified staple food crops under Harvestplus. Afr J Fd Agric Nutr Dev. 17:11905–35.

Andhra Pradesh cuisine. Indianfoodforever. com. Retrieved 4 June, 2012.

Anita M. 2004. Culinary fictions: Food in South Asian diasporic culture. Temple Univer Press. 233. ISBN 978-1-4399-0078-9. Retrieved, 2012.

Ann G L, Nikki A F, Frank BHu, Kathleen MZ, Dariush M and Penny M. 2017. Kris-Etherton Nutr J. 16: 53.

Anyawu C U and Nwosu G C. 2014. Assessment of the antimicrobial activity of aqueous and ethanolic extract of Piper guineense leaves. J Med Res. 8:337–439.

Aranceta J and Pérez R C. 2012. Recommended dietary reference intakes, nutritional goals and dietary guidelines for fat and fatty acids: a systematic review. Brit J Nutr. 107(S2): S8-S22.

Aranceta J and Serra Majem L. 2011. Grupo Colaborativo para la actualizacio´n de los Objetivos Nutricionales para la Poblacio´n Espanola. Consenso de la Sociedad Espanola de Nutricion Comunitaria.

Aruna T and Arlene B. 2012. Multicultural Handbook of Food, Nutrition and Dietetics. John Wiley an Sons. 5. ISBN 978-1-118-35046-1. Retrieved 2012.

Arunachal Pradesh food. 2013. Arunachal Pradesh. ifood. tv. Archived from the original on 24 April. Retrieved, 2012.

Arvind K and Pushpalata D. 2006. Green Technologies for Sustainable Agriculture. Daya Publishing House. 28. ISBN 978-81-7035-419-2. Retrieved 2012.

Ashok P K and Upadhyaya K. 2012. Tannins are asrigent. J Pharmac Phytochem. 1:45–50. asianinspirations. com. au.

Assam General Knowledge. Bright Publications. 79. ISBN 978-81-7199-451-9. Retrieved 2012.

Astrup A, Dyerberg J, Elwood P, et al. 2011. The role of reducing intakes of saturated fat in the prevention of cardiovascular disease: where does the evidence stand in 2010? The Amer J clin Nutr. 93(4): 684-688.

Babu S, Gajanan SN, Sanyal and Prabuddha. 2014. Food security, poverty and nutrition policy analysis: Statistical Methods and Applications: Sec Ed. Food Security, Poverty and Nutrition Policy Analysis: Statistical Methods and Applications: Second Ed. 1-615.

Bai Chao and Twyman, Richard and Farré, Gemma and Sanahuja, Georgina and Christou, Paul and Capell, Teresa and Zhu, Changfu. 2011. A golden era pro-vitamin A enhancement in diverse crops. *In Vitro* Cellular Develop Biol Plant. 47. 205-221. 10. 1007/s11627-011-9363-66.

Baker et al. 1999. Iron fortification of infant formulas. Pediatrics. 194:1. http://pediatrics. aappublications. org/content/104/1/119.

Balasubramanian D. 2008. Potato: historically important vegetable. The Hindu. Chennai, India. Retrieved 2012.

Balié J. 2009. Methodological Guide for Policy Intelligence and Preparedness work in pilot countries. Draft. Rome: FAO.

Barlow SE. 2007. Expert committee recommendations regarding the prevention, assessment and treatment of child and adolescent overweight and obesity: summary report. Pediatrics 120(Suppl. 4):S164–92.

Baron JA, Beach M, Mandel JS, van Stolk RU, Haile RW, Sandler RS, et. al. 1999. Calcium supplements for the prevention of colorectal adenomas. N Engl J Med: 340:101-7.

Barona J and Fernandez M. 2012. Dietary cholesterol affects plasma lipid levels, the intravascular processing of lipoproteins and reverse cholesterol transport without increasing the risk for heart disease. Nutrients - Open Access Hum Nutr J. 4.

Barrena R and Sánchez M. 2009. Differences in consumer abstraction levels as a function of risk perception. J Agric Eco. 61(1):2010:32-99.

Basappa S, Somashekar D, Agrawal Renu, Suma K and Bharathi K. 2009. Nutritional composition of fermented ragi (chhang) by phab and defined starter cultures as compared to unfermented ragi (*Eleucine coracana* G.). Intern J Fd Sci Nutr. 48. 313-319. 10. 3109/09637489709028577.

Basith S, Cui M, Hong S and Choi S. 2016. Harnessing the therapeutic potential of capsaicin and its analogues in pain and other diseases. Molecules. 21:(8) 966. Epub 2016, Jul 23.

Bassey MW and Schmidt OG. 1989. Abrasive-disk dehullers in africa, from research to dissemination. Inter Develop Res Cent (IDRC):Ottawa, Canada. 1(1):27-35.

Baysal A, Merdol TK, Tasci NC, Sacir FH and Basoglu S. 2006. Synthesis report No 6: Traditional Foods in Europe Dr. Elisabeth Weichselbaum and Bridget Benelam British Nutrition Foundation Dr. Helena Soares Costa National Institute of Health (INSA), Portugal). Samples from Turkish cuisine. Food Safety Department, Community Nutr Div, Ministry of Health, Republic of Turkey, Ankara. p54.

Bearth Angela, Cousin Marie-Eve and Siegrist Michael. 2014. The consumer's perception of artificial food additives:Influences on acceptance, risk and benefit perceptions. Food Quality and Preference. 38. 14–23. 10. 1016/j. foodqual. 2014. 05. 008.

Beckeman M and Skjöldebrand C. 2007. Clusters/networks promote food innovations. JFd Eng. 79. 1418-1425. 10. 1016/j. jfoodeng. 2006. 04. 024.

Bendsen NT, Christensen R, Bartels EM, et al. 2011. Consumption of industrial and ruminant trans fatty acids and risk of coronary heart disease: a systematic review and meta-analysis of cohort studies. Eur J clin nutr. 65(7): 773-78.

Benjamin U, Ani O, Abada, U, Ugwuishiwu B, Jude O and Nwakaire J. 2016. Small-scale food processing enterprises: measures for national development and addressing food security challenges in Nigeria. Intern J Sci Tech Res Eng. 1. 72-82.

Berk Z. 1992. Technology of production of edible flours and protein products from soybeans. FAO Agricultural Services Bulletin No. 97. FAO of UN, Rome.

Besch M. 2002. Globalisierung und Regionalisierung in der Ernährung – Fast Food versus Slow Food. In: Ernährung und Raum: Regionale und ethnische Ernährungsweisen in Deutschland. Gedrich K, Oltersdorf U(eds), Bundesforschungsanstalt für Ernährung, Karlsruhe.

Besong E, Balogun M and Djobissie SFA, Obu D and Obimma J. 2016. Medicinal and economic value of Dialium guineense. 19: 163-170.

Bevan S and Szolcsányi J. 1990. Sensory neuron—specific actions of capsaicin: mechanisms and application. Trends Pharmacol Sci. 11: 330– 333.

Bhagavad-Gita. 2013. Chapter 17; Verse 8, 9, 10.

Bhandari L. 2009. Indian States at a Glance 2008–09: Performance, Facts And Figures–North-East And Sikkim. Pearson Education India. 48. ISBN 978-81-317-2348-7. Retrieved, 2012.

Bhandari L. 2009a. Indian States at a Glance 2008–09: Performance, Facts And Figures – Bihar. Pearson Education India. 30. ISBN 978-81-317-2333-3. Retrieved 2012.

Bhandari L. 2009b. Indian States at a Glance 2008–09: Performance, Facts and Figures – West Bengal. Pearson Education India. 30. ISBN 978-81-317-2352-4. Retrieved, 2012.

Bhandari L. 2009c. Indian States at a Glance 2008–09: Performance, Facts and Figures – Kerala. Pearson Education India. 36. ISBN 978-81-317-2340-1. Retrieved, 2012.

Bhasin K. 2012. This is the difference between invention and innovation. Business Insider. www. businessinsider. com/this-is-the-difference-between-invention-and-innovation-2012-4.

Bhat, Shripad. 2012. Importance of traditional food system, in recent trends in food science and technology. P Sadananda Maiya (Ed). Tumkur University, 2012. ISBN: 9789382694007.

Bhutta Z A, Bawany F A, Feroze A and Rizvi A. 2009. The Impact of the food and economic crisis on child health. Draft Working Paper prepared for UNICEF Conference, East Asia and the Pacific Islands. Singapore. 6–7 January.

Bickel G, Nord M, Price C, Hamilton W and Cook J. 2000. Guide to measuring household food security. USDA Food and Nutrition Service Caballero.

Birkland T. 2010. An Introduction to the Policy Process: Theories, Concepts, and Models of Public Policy Making. M. E Sharpe Inc. 3rd edition. Committee on World Food Security (CFS). 2015. Global Strategic Framework for Food Security and Nutrition: Fourth Version.

Black RE, Victora CG, Walker SP, Bhutta ZA, Christian P, de Onis M, Ezzati M, Grantham-McGregor S, Katz J, Martorell R and Uauy R. 2013. Maternal and child undernutrition and overweight in low-income and middle income countries. The Lancet. 382(9890): 427–451.

Blancquaert D, Van Daele J, Strobbe S and Kiekens F. 2015. Improving folate (vitamin B9) stability in biofortified rice through metabolic engineering. Nat Biotechnol. 33(10):1076–8.

Booth D and Cammack D. 2013. Governance for Development in Africa: Solving Collective Action Problems. London: Zed Books.

Booth D and Golooba-Mutebi F. 2012. Developmental Patrimonialism? The Case of Rwanda. African Affairs. 111(444): 379-403.

Booth D, Dietz T, Golooba-Mutebi F, Fuady A H, Henley D, Kelsall T, van Donge J K. 2015. Developmental Regimes in Africa: initiating and sustaining developmental regimes in Africa. Synth Rep.

Booth D. 2012. Development as a collective action problem: Addressing the real challenges of African governance. lLondon:ODI (Overseas Development Institute)Africa Power and Politics Programme.

Booth D. 2014. Towards a developmental state concept that is fit for purpose (in the 21st century, in Africa). Paper presented at the African Studies Association Annual Meeting, Indianapolis 20-22 November.

Bordes J, Branlard G, Oury FX, Charmet G, Balfourier F. 2008. Agronomic characteristics, grain quality and flour rheology of 372 bread wheat's in a worldwide core collection. J Cer Sci. 48: 569-579.

Borneo Ediz Inglese. 2008. Lonely Planet. 97. ISBN 978-1-74059-105-8.

Bosco Bua and Christopher Onang. 2017. Validating the role of african indigenous vegetables for food and nutrition security in Uganda. J Fd Sci Eng. 7:316-322 doi: 10. 17265/2159-5828/2017. 06. 005.

Bostick RM, Kushi LH, Wu Y, Meyer KA, Sellers TA and Folsom AR. 1999. Relation of calcium, vitamin D and dairy food intake to ischemic heart disease mortality among postmenopausal women. Am J Epidemiol. 149(2):151-61.

Bouis H and Islam Y. 2012. Biofortification, chapter 10. In: Fan S, Pandya-Lorch R, editors. Leveraging agriculture to reduce hidden hunger in reshaping agriculture for nutrition and health(IFPRI2020 book). https://www. ifpri. org/publication/reshapingagriculture-nutrition-and-health.

Bouis H. 2018. Reducing mineral and vitamin deficiencies through biofortification:Progress under HarvestPlus. In:Biesalski HK, Birner R, eds. Hidden hunger:Strategies to improve nutrition quality, world review of nutrition and dietetics. Basel: Karger. 12–22.

Breiman A and Graur D. 1995. Wheat Evolution. Israel J Pl Sci. 43: 85-98.

Brenna T, Salem N, Sinclair A, et al. 2009. ISSFAL official statement number 5, α-linolenic acid supplementation and conversion to n-3 long chain PUFA in humans. Prostaglandins, Leukotrienes and Essential Fatty Acids 80(2/3): 85-91.

Brian Krans. 2016. The Grapefruit Diet what Is It?In Healthline, rev. by Natalie Butler.

Brookes G and Barfoot P. 2016. GM crops: global socio-economic and environmental impacts. 1996–2014. www. pgeconomics. co. uk/pdf/2016globalimpactstudymay2016. pdf.

Brouwer I, Wanders A and Katan M. 2013. Trans fatty acids and cardiovascular health: research completed? Eur J Clin Nutr. 67(5): 541-547.

Brown KH, Rivera JA, Bhutta Z, Gibson RS, King JC, et al. 2004. International Zinc Nutrition Consultative Group (IZiNCG) technical document#1. Assessment of the risk of zinc deficiency in populations and options for its control. Fd Nutr Bull. 25(1 Suppl 2):S99–203.

Browning KN and Travagli RA. 2014. Central nervous system control of gastrointestinal motility and secretion and modulation of gastrointestinal functions. Compr Physiol. 4(4): 1339–1368. doi:10. 1002/cphy. c130055

Bruce F J. 1958. The Staple Food Economies of Western Tropical Africa. Stanford University Press. 14. ISBN 978-0-8047-0537-0. retrieved 2012.

Brzozowski T, Konturek SJ, Sliwowski Z, Pytko-Polończyk J, Szlachcic A and Drozdowicz D. 1996. Role of capsaicin-sensitive sensory nerves in gastroprotection against acid-independent and acid-dependent ulcerogens. Digestion. 57(6):424-32.

Buisness world, April, 2020. Garhwal and Kumaon Cuisine.

Caballero PA, Gomez M and Rosell CM. 2007. Improvement of dough rheology, bread quality and bread shelf-life by enzymes combination. J Fd Eng. 81: 42–53.

Calcutta Cuisine - Cuisine Of Kolkata India - Bengali Traditional Food - What to eat in Calcutta India. www. kolkata. org. uk. Retrieved 2019.

Care for some Sindhi savouries?2010. Dnaindia. com. Retrieved 7 February 2011.

CELAC, 2017. Governance of food and nutrition security Factors for viability and sustainability Case studies from seven Latin American countries.

Chandigarh Cuisine.

Chandrasekhar K, Sreevani S, Seshapani P and Pramodhakumari J. 2012. A Review on palm wine. Internl J Res Biol Sci. 2 (1): 33-38.
Chapman, P. 2009. India: food & cooking: The ultimate book on Indian cuisine. New Holland Publishers. 38–39. ISBN 978-1-84537-619-2. Retrieved 7 July 2012.
Chapter 17, Verse 8, 9, 10. Bhagavad-Gita. Retrieved 2011.
Charlotte D and Marie-Caroline Dodé. 2016. Nutrition and Food systems Division, FAO learning LAB, Agriculture, Nutrition and Health Academy Week.
Chaudhuri UN, Burnet RB, Kirkham MB and Kanematsu ET. 1986. Effect of carbon dioxide on sorghum yield, root growth and water use. Agri Forest Meteor. 37: 109-122.
Chhattisgarh Cuisine in India. 2005. India9. com. Retrieved 2011. Mahua flower liquor. Fieldsite in Central India. 2012.
Chhattisgarh delicacies. Chhattisgarh culture. journeymart. com. Retrieved 2012.
Christian P. 2010. Impact of the economic crisis and increase in food prices on child mortality: Exploring nutritional pathways. The JNutr. 140(1):177S–181S. DOI:10. 3945/ jn. 109. 111708.
Chromiak JA and Antonio J. 2002. Use of amino acids as growth hormone-releasing agents by athletes. Nutri. 18: 657-661.
Chung et al., 2011. Federal reserve bank of san francisco working paper series. Have we underestimated the likelihood and severity of zero lower bound events? Working Paper2011-01 http://www. frbsf. org/publications/economics/papers/2011/wp11-01bk. pdf JMCB44(s1):47-82. https://doi. org/10. 1111/j. 1538-4616. 2011. 00478. x
Clifford A W. 2005. Some like it hot: spicy favorites from the world's hot zones. Harvard Common Press. 236. ISBN 978-1-55832-269-1. Retrieved 2012.
Colleen TS. 2004. Food culture in India. Greenwood Publishing Group. 94. ISBN 978-0-313-32487-1. Retrieved 2012.
Commission on Social Determinants of Health (CSDH). 2008. Closing the gap in a generation: health equity through action on the social determinants of health. Geneva, Switzerland, WHO.
Conseil Supérieur de la Santé. 2009. Publication Du Conseil Superieur De La Santé N° 8309. Recommendations nutritionnelles pour la Belgique, Révision 2009. Bruxelles, Belgique.
Cornell HJ and Hoveling AW. 1998. Wheat:Chemistry and Utilization. Technomic Publishing Company. Lancaster. 43-53.
Costa. 2005. Traditional Foods in Europe - EuroFIR www. eurofir. org.
Cowed down by the ban. 2012. Hindustan Times. Archived from the original on 20 February, 2012.
Crider K, Bailey L and Berry R. 2011. Folic acid food fortification—its history, effect, concerns and future directions. Nutrients. 3(3):370–84.
Cuisine of Kerala. Retrieved, 2016.
Cuisine of Kerala. 2016. Zero oil south indian cook book. Dr. Bimal Chhajer. ISBN 978-81-288-0512-7.
Cuisine. 2012. India cuisine. Incredible India. Archived from the original on 2 June Retrieved,.
Cuisines of Andaman and Nicobar Islands. 2012. Andaman and Nicobar Islands. indfy. com. Retrieved , 2012.
Cuisines of Andaman and Nicobar Islands. 2012. Andaman and Nicobar Islands. indfy. com. Retrieved, 2012.
Cuisines of Delhi, famous Delhi Cuisine, famous Delhi food, famous food in Delhi. 2013. Indiasite. Retrieved, 2013.
DAC Network on gender equality by Justina Demetriades. 2009. Based on BRIDGE's Gender and Indicators Cutting Edge Pack. 2007. http://www. bridge. ids. ac. uk/reports_gend_CEP. html#Indicators

Dalal T. 2010. p7. ISBN : 978-8-189491-72-7. Copyright © Sanjay & Co. "Tarla Dalai" is also a registered trademark owned by Sanjay & Co.

David M. 2018. Review of the role of political economy with respect to agricultural policy making in Africa. Int. J. Sci. Res. SH-2018-171-179 DOI https://doi. org/10. 18535/ijsrm/ v6i6. sh03.

Davidson A. 2014. The Oxford Companion to Food (3rd ed.). Tom Jaine (ed.). Oxford: Oxford Univ Press. ISBN 0-19-967733-6. 21–22.

De la Peña I, Garrett J and Gelli A. 2018. Nutrition-sensitive value chains from a smallholder perspective: a framework for project design. Rome, IFAD

de Onis M and Branca F. 2016. Childhood stunting:a global perspective. Maternal and Child Nutr. 12(Suppl. 1): 12–26.

de Schutter O. 2014. Food for thought: learning from Brazil. A Source of inspiration for the successful implementation of 'Zero Hunger' in West Africa. London, Oxfam International. United Nations Economic Commission for Africa. The Cost of Hunger in Africa: Social and Economic Impact of Child Undernutrition in Egypt, Ethiopia, Swaziland and Uganda. Report.

Delish potpourri. 2016. www. delishpotpourri. com

Demetri J and Nascimbeni P. 2008. Italian regional food. Online: http://www. lifeinitaly. com/ food/italian-regional-food. asp (accessed 24 September, 2008).

Department of economic and social affairs office for ecosoc support and coordination achieving sustainable development and promoting development cooperation dialogues at the economic and social council United Nations, New York. 2008.

Devagi S. 2007. Naturally Speaking: Indian: Recipes and Home Remedies. Marshall Cavendish. 69. ISBN 978-981-232-715-4. Retrieved 2012.

Devarajan, Agilandeswari & Mohanmarugaraja, M. (2017). A Comprehensive Review on Rasam: A South Indian Traditional Functional Food. Pharmacognosy Reviews. 11. 73. 10. 4103/phrev. phrev_13_17.

Dewan M L, Karan S and Piyoosh R. 2001. People's Movement for Himalayan Rejuvenation. Concept Publishing Company. 93. ISBN 978-81-7022-815-8. Retrieved 2012.

DFID (Department For International Development). 2006. Evaluation Report 683. Country programme evaluations synthesis of 2006/07. Evaluations by Barr J and Vaillant C, ITAD. Jan. 2008.

Diane E T, Darren C G, Charlotte E L E, Christine L C, Camilla N, Charlotte W, Janet E C, Christopher P G and Victoria J B. 2013. Dietary fibre intake and risk of cardiovascular disease:systematic review and meta-analysis. BMJ : 347. doi: https://doi. org/10. 1136/ bmj. f6879 (Published 19 December 2013).

Dias. 1996. Steward, The orient blackswan. 215. ISBN 978-81-250-0325-0. Retrieved, 2012.

Dietary Guidelines for Americans. 2015-2020. health. gov

Dietary reference intakes for calcium and vitamin D. Washington, DC: National Academy Press. 2010. Office of dietary supplements.

Dike M C and Ahamefula N E. 2012. Comparative study of proximate, phytochemical and mineral compositions of edible plant fruits/seeds from Nigerian rainforest. Intern J Biol Chem Sci. 6:1905–1909.

Dinesh S and Madalsa S. 2014. The Science and Art of Indian Cooking: Indian Cooking. 20. ISBN 978-1-4828-2215-1.

Diouf M, Gueye M, Faye B, Dieme O and Lo C. 2007. The commodity systems of four indigenous leafy vegetables in senegal. Water SA 33: 343-8.

Djoussé L and Gaziano JM. 2009. Dietary cholesterol and coronary artery disease: a systematic review. Curr Atheroscl Rep. 11(6):418-422. doi: 10. 4103/ijmr. IJMR_1893_18.

Dolui A K and Segupta R. 2012. Antihyperglycemic effect of different extracts of leaves of Cajanus cajan HPLC profile of the active extracts. Asian J Pharma Clin Res. 5.

Drewnowski A and Popkin BM. 1997. The Nutrition Transition: New Trends in the Global Diet. Nutri Rev. 55(2):31-43.

Dubé L, Addy NA, Blouin C and Drager N. 2014. From policy coherence to 21st century convergence: a whole-of-society paradigm of human and economic development. Ann N Y Acad Sci. 1331:201-215. doi: 10. 1111/nyas. 12511. Epub 2014 Aug 2.

Dubey Krishna gopal. 2011. The Indian cuisine. PHI Learning. 233. ISBN 978-81-203-4170-8. Retrieved, 2012.

Dubock A. 2013. Golden Rice: a long-running story at the watershed of the GM debate. 1–12. http://b4fa. org/wp-content/uploads/2013/10/Viewpoints-Dubock. pdf and[printable]. http://www. goldenrice. org/PDFs/GR_A_long-running_story. pdf. Accessed 25 Sept 2017.

Dubock A. 2014a. The politics of golden rice. GM Crop Food. 5(3):210–22.

Dubock A. 2014b. The present status of Golden Rice. J Huazhong Agric Univ. 33(6):69. http:// www. goldenrice. org/PDFs/DubockThe_present_status_of_Golden_Rice-2014. pdf.

Dubock AC. 2013. Nutritional enhancement by biofortification of staple crops, chapter 12. In: Bennett DJ, Jennings RC(Ed.). Successful agricultural innovation in emerging economies: new genetic technologies for global food production. Cambridge: Cambridge University Press. 119–220.

Dubock AC. 2017. Biofortified Golden Rice as additional intervention for vitamin A deficiency, chapter 32. In: Sasaki T, Ed. Achieving sustainable cultivation of rice: breeding for higher quality. Cambridge: Burleigh Dodds Science Publishing Limited. 1.

Dum Pukht cooking. 2012. Compendium of food terms. theflavoursofhistory. com. Archived; 2 April, 2012. Retrieved, 2012.

Ebbert J and Jensen M. 2013. Fat depots, free fatty Acids, and dyslipidemia. nutrients 5(2): 498-508.

Echo IA, Osuagwu A N, Agbor R B, OkakaE C and Ekanem B E. 2012. Phytocemical composition of Afromomum melegueta and Piper guineense seeds. World J Appl Eviron Chem. 2:17–21.

Edelstein S. 2011. Food, Cuisine, and Cultural Competency for Culinary, Hospitality and Nutrition Professionals. Jones and Bartlett Publishers. 276. ISBN 978-1-4496-1811-7. Retrieved, 2012.

Edgar T and Showick T. ed. 2010. The pearson concise general knowledge manual (New ed.). Pearson Education India. 34. ISBN 978-81-317-2766-9.

Ekanem A P, UdohV F and Oku E E. 2010. Effects of ethanol extract of Piper guineense seeds (Schum and Thonn) on the conception of mice (Mua Musculus)Afr J Pharm Pharmac. 4:362–367.

Ekwenye U N and Okorie C F. 2010. Antibacterial activity of Tetrapleura Tetraptera Taub. Pod extracts. Intern J Pharm Biol Sci. 1: 71–78.

El kalif AE and E Tinay AH. 2002. Effect of cysteine on bakery products from wheat sorghum blends. Food chem. 77:133-137.

El Rasheed Ahmed Salim. 2017. Fortified sorghum as a potential for food security in rural areas by adaptation of technology and innovation in Sudan. JFd Nutr. Popul Health. 1:1.

Elmadfa I. 2009. European Nutrition and Health Report 2009. 62. Vienna, Austria.

Elsner A E , Burns S A, Beausencourt E and Weiter J J. 1998. Foveal cone photopigment distribution: small alterations associated with macular pigment distribution. 39(12):2394-2404. doi:https://doi. org/

Emir TLR. 2017. Ed. Boca Raton (FL, 2nd ed.): CRC Press/Taylor and Francis, Neurobiology of TRP Channels. 2nd edition. Neurobiology of TRP Channels.

Erber E, Beck L, De Roose E et al. 2010. Prevalence and risk factors for self reported chronic disease amongst inuvialuit populations. J Hum Nutr Diet. 23: 43–50. eal challenges of Aan governance

Esteves and Lesley A. 2008. Food court Nawabi Bhopal is a long and large meal. Outlook traveller. Archived from the original on 31 January 2013. Retrieved , 2012.
Etim OE, Chinaka NS and Dueu R U. 2014. Effect of Piper guineense and Azadarica indica on some polycyclic aromatic hydrocarbons (PAH) levels of germinating Telferia occidentalis (ugu). Intern J Environ Bioremed Biodegrad. 2: 151–159.
EUFIC website, Energy Balance section. Accessed, 4 December 2014.
Eugene F and William C. 1964. Fodor's guide to India. D. McKay. Retrieved, 2012.
European Commission DG Joint Research Centre. 2013. JRC Science and Policy Reports:Trans fatty acids in diets: health and legislative implications. european union, Luxembourg.
European Commission, Joint Research Centre, Institute for Health and Consumer Protection. 2011. Nutrition Research Highlights 2. Retrieved from Nutrition Research Highlights, 2011.
European Commission. FAO. 1996.
European Commission. 2007a. European policy for quality agricultural products. Fact sheet. Office for Official Publications of the European Communities, Luxembourg.
European Commission. 2007b. European Research on Traditional Foods. Community Research. Office for Official Publications of the European Communities, Luxembourg.
European Food Safety Authority (EFSA). 2010 a. Scientific Opinion. EFSA Journal 8(3):1461.
European Food Safety Authority (EFSA). 2010 b. Scientific Opinion. EFSA Journal 8(10): 1734.
European Food Safety Authority (EFSA). 2011. Scientific Opinion. EFSA J 9(5):2168.
European Union (EU). 2013. Enhancing Maternal and Child Nutrition in External Assistance: an EU Policy Framework. Communication from the Commission to the European Parliament and the Council. Brussels, European Commission. FAO. 1996.
FAO and FHI 360. 2016. Minimum Dietary Diversity for Women- A Guide to Measurement. http://www. fao. org/3/a-i5486e. pdf
FAO Focus: Women and food security: women hold the key to food security. Rome. 2009. Securing sustainable small-scale fisheries:Bringing together responsible fisheries and social development. In report of the global conference on small-scale fisheries. Bangkok. 2010. The State of Food and Agriculture 2010.
FAO learning LAB. 2016. Mapping and analysis policies to inform food security and nutrition planning:Towards more coherent and cross-sectoral policies charlotte dufour and marie-caroline dodé nutrition and food systems division, FAO learning LAB, 2016. Agriculture, Nutrition and Health Academy Week 2016.
FAO of UN. 1992. The international conference on nutrition. http://www. fao. org/docrep/v7700t/v7700t02. htm.
FAO report. 2009.
FAO SOFA. 2013. Food systems for better nutrition.
FAO. 1996.
FAO. 2002.
FAO. 2008. An introduction to the basic concepts of food security. Rome, Italy:FAO. Retrieved; http://www. fao. org/docrep/013/al936e/al936e00. pdf
FAO. 2010. http://www. fao. org/3/a-i1951e. pdf.
FAO. 2011.
FAO. 2012.
FAO. 2013.
FAO. 2015.
FAO. 2018. ISBN 978-92-5-131212-4 (FAO) © FAO, 2018). Leveraging Small and Medium Enterprises to improve nutrition; Published by the Food and Agriculture Organization of the United Nations and the Global Alliance for Improved Nutrition, Rome.
FAO. 1995. Annex-Some recipes based on sorghum and millets. Sorghum and millets in human nutrition. FAO corporate document perpository. FAO, Food and Nutrition Series. No. 27.

FAO. 1996. Rome declaration on world food security and world food summit plan of action.

FAO. 2001. Crop and food supply situation in Sudan. Special Report.

FAO. 2006. Livestock's long shadow environmental issues and options. Rome, Italy:FAO. Retrieved; ftp://ftp. fao. org/docrep/fao/010/a0701e/a0701e. pdf

FAO. 2013. Food Security Status of U. S. Households in 2012. USDA-ERS.

FAO/UN. 2018. Achieving Zero Hunger in Africa by 2025. African union high-level meeting, Addis Ababa, Ethiopia.

FAO/WHO and 2000, 2010.

FAO/WHO. 2014.

Farvid MS, Ding M, Pan A et al. 2014. Dietary linoleic acid and risk of coronary heart disease: a systematic review and meta-analysis of prospective cohort studies. Circulation. 130:1568-1578.

Fattore E and Fanelli R. 2013. Palm oil and palmitic acid: a review on cardiovascular effects and carcinogenicity. Intern J Fd Sci Nutr. 64(5):648-659.

Feed your 'Desi Mania' at Nirula's". Fnbnews. 2010. Retrieved, 2012.

Feleke H, Srinivasulu A, Kanikella S, Aruna B Biswas, Jaganmoy SM, Rao A and Narayana PV. 2018. Energy Dispersive X-Ray fluorescence elemental analysis of roasted and non-roasted ethiopian coffee specialty. 12. 51-70. 10. 9790/2402-1204015170.

Finkelstein JL, Mehta S, Udipi SA, Ghugre PS, Luna SV, Wenger MJ, et al. 2015. A randomized trial of iron-biofortified pearl millet in school children in India. J Nutr. 145:1576–81.

Finney KF, Yamazaki WT, Youngs VL and Rubenthaler GL. 1987. Quality of hard, soft, and durum wheats. Wheat and wheat improvement-agronomy monograph. 13: 677-748.

Flavours of Maharashtra at Renaissance. 2010. The Times of India. 12 July. Retrieved, 2012.

Fletcher J. 2017. All you need to know about fat-soluble vitamins. Rev. Debra Rose Wilson.

Flyman M and Afolayan A. 2006. The suitability of wild vegetables for alleviating human dietary deficiencies. S Afr J Bot. 72: 492–497.

Food and Agriculture Organization of the United Nations (FAO) and World Health Organization (WHO). 2010. http://www. fao. org/3/a-i1951e. pdf

Food and Agriculture Organization of the United Nations, International Fund for Agricultural Development and World Food Programme. 2015. The state of food insecurity in the world. 2015. Meeting the 2015 International Hunger Targets: Taking stock of uneven progress. Rome: FAO.

Food and Agriculture Organization of the United Nations. 2009. Draft declaration of the world summit on food security. World Summit on Food Security, Rome, 16–18 November 2009. Rome: FAO.

Food and Agriculture Organization of the United Nations. 2012. Traditional Foods:Better Livelihoods, Dietary Diversity and Health. Rome:FAO. http://teca. fao. org/technology/traditional-foods-better-livelihoods-dietarydiversity-and-health

Food Habits - Banglapedia. en. banglapedia. org. Retrieved, 2019.

Food in Mizoram. Mapsofindia. com. 2013. Retrieved, 2013.

Food of Haryana, Cuisine of Haryana, Famous Haryana Food, Recipes of Haryana, Haryana Food. Richindianculture. com. Retrieved, 2011.

Forouzanfar M H, Lily A, Anderson H R, Bachman V F, Biryukov S et al. 2015. Epub 2015. Global, regional, and national comparative risk assessment of 79 behavioural, environmental and occupational, and metabolic risks or

Fortification of Foods Regulations. 2018. India: Food Fortification Laws, 2018- 19 September 2018. S. S. Rana & Co. Advocates.

Freedman R. L. 2015. Author of indigenous wild food plants in home gardens: improving health and income - with the assistance of agricultural extension. Intern J Agri Exten. 3:1.

Frewer L J, Howard C, Hedderley D and Shepherd R. 1997. Methodological approaches to assessing risk perceptions associated with food-related hazards. Risk Analysis. 18(1): 95-102.

FSSAI. 2016. The principal regulations were published in the Gazette of India, Extraordinary vide notification number F. No. 2-15015/30/2010, dated the 1st August, 2011.

G20 Ministers. 2016. G20 agriculture ministers meeting Communiqué. http://www. g20. utoronto. ca/2016/160603-agriculture. html.

Galiè N, Teufel L K, BaltenweckI, Webb AGirard, Dominguez-Salas P and Yount KM. 2018. The women's empowerment in livestock index. Soc. Indicat. Res. 142 (2):799-825.

Galvin KA , Beeton TA, Boone RB and BurnSilver SB. 2015. Nutritional status of Maasai pastoralists under change. Hum. Ecol. 43 (3): 411-424. https://doi. org/10. 1007/s10745-015-9749-x

Game cuisine: A Rajput legacy by Madhulika Dash. 2014

Gannon B, Kaliwile C, Arscott SA, et al. 2014. Biofortified orange maize is as efficacious as a vitamin A supplement in Zambian children even in the presence of high liver reserves of vitamin A:a community-based, randomized placebo-controlled trial. AmJClinNutr. 100(6):1541–1550. doi:10. 3945/ajcn. 114. 087379.

Gavrilova R. 2005. Bulgaria. Golden fruits from the orchards. In:Culinary cultures of Europe. Identity, diversity and dialogue (Goldstein C and Merkle K, eds). Council of Europe Publishing, Verlagsgruppe Lübbe, Germany. 95-102.

Gender equality and food security. 2013. Gender equality and food security—women's empowerment as a tool against hunger Mandaluyong City, Philippines:Asian Development Bank. Women's Empowerment as a Tool against Hunger © 2013 Asian Development Bank. Printed in the Philippines. ISBN 978-92-9254-171-2 (Print), 978-92-9254-172-9 (PDF) Publication Stock No. RPT135782-3 Cataloging-In-Publication Data Asian Development Bank. 2013.

Global food policy report. 2018. International food policy research institute (IFPRI)global food policy report. 2018. FPRI's flagship report.

Global food policy report. 2016. Washington, DC: IFPRI; International Food Policy Research Institute.

Global Nutrition Report. 2017. Nourishing the SDGs.

Global Panel. 2016. Global Panel on Agriculture and Food Systems for Nutrition Working Paper. The economics of reducing malnutrition in Sub-Saharan Africa1 John Hoddinott Cornell University.

Goan cuisine. 2012. Goa dining guides. Archived from the original on 10 July Retrieved, 2012.

Gomes, T. ; Feas, X. ; Iglesias, A. ; Estevinho, L. M. 2011. Study of Organic Honey from the Northeast Portugal. Molecules, 16, 5374–5386.

Gómez-Galera S, Rojas E, Sudhakar D, Zhu C, Pelacho AM, Capell T, et al. 2010. Critical evaluation of strategies for mineral fortification of staple food crops. ransgenic Res. 19:165–80.

Govindnarayana VS and Satyanarayana MN. 1991. Capsicum production, technology-chemistry and quality part V. Impact on physiology, pharmacology, nutrition and metabolism. Crit Rev Fd Sci Nutr. 29: 437–474.

Gregory T and Sewando P. 2013. Determinants of the probability of adopting of Quality Protein Maize (QPM) technology in Tanzania: A logistic regression analysis. Int J Dev Sustain. 2:729–46.

Guenther P M, Reedy J and Krebs-Smith SM. 2008a. Development of the healthy eating index-2005. J Amer Diet Ass. 108 (11):1896–1901.

Guenther P M, Reedy J and Krebs-Smith SM. 2008b. Global Panel on Agriculture and Food Systems for Nutrition. London, UK.

Gunaratna NS, De Groote H, Nestel P, Pixley KV and McCabe GP. 2010. A meta-analysis of community-level studies on quality protein maize. Fd Policy. 35:202–10.

Gupta HS, Hossain F, Muthusamy V. 2015. Biofortification of maize:An Indian perspective. Indian J Genet Plant Breed. 75:1–22.

Gupta NT, Sharma and Vishwanath. 2016. Bioavailability of Micronutrients in Crops for better human health. 10. 13140/RG. 2. 2. 36759. 16801.

Hamlet B. 2001. Encyclopaedia of North-East India: Meghalaya. Mittal Publications. 47. ISBN 978-81-7099-791-7. Retrieved, 2012.

Haniadka R S, Elroy S, Venkatesh P, Princy F, Raja B and Shrinath. 2013. A review of the gastroprotective effects of ginger (Zingiber officinale Roscoe). Fd Funct. 4. 10. 1039/ c3fo30337c.

Harika RK, Eilander A, Alssema M, et al. 2013. Intake of fatty acids in general populations worldwide does not meet dietary recommendations to prevent coronary heart disease: a systematic review of data from 40 countries. Annals of Nutrition and Metabolism.

Harvey T, Zkik A, Auges M and Clavel T. 2016. Assessment of iron deficiency and anemia in pregnant women:an observational French study. Womens Health (Lond). 12(1):95–102. doi:10. 2217/whe. 15. 91

Hawkes C. 2015. Enhancing coherence between trade policy and nutrition action: implementing the framework for action of the second international conference on nutrition. Discussion paper 1. United Nations Standing Committee on Nutrition, Geneva.

Hellin J and Erenstein O. 2009. Maize poultry value chains in India: implications for research and development. J New Seeds. 10:245–69.

Henrietta A O, Chidozie G U and Ada E KO. 2017. The role of spices in nutrition and health: a review of three popular spices used in Southern Nigeria. Food Quality and Safety. 1(3): 171–185. https://doi. org/10. 1093/fqsafe/fyx020.

Herforth A and Ahmed S. 2015. The food environment, its effects on dietary consumption and potential for measurement within agriculture nutrition interventions. Fd Sec. 7(3):505-520.

Herforth A. 2015. Access to adequate nutritious food:New indicators to track progress and inform action. In: Sahn, D (ed.):The fight against hunger and malnutrition. Oxford University Press.

Herforth A. 2016. What is needed for assessing the impact of food systems policies on food environments and diets? Implementing the framework for action of the second international conference on nutrition. UN Standing Committee on Nutrition.

Herrera C, Grossman J B, Kauh T J, Feldman A F and McMaken J. (with Jucovy, L. Z.). 2007. Making a difference in schools: The Big Brothers Big Sisters school-based mentoring impact study. Philadelphia, PA: Public/Private Ventures.

High C and Shackleton C M. 2000. The comparative value of wild and domestic plants in home gardens of a south african rural village. Agroforestry Sys. 48 (2): 141-56.

High time to make most of mahua – Times of India. The Times of India. Retrieved, 2018.

Highlights of Agriculture Census. 2010-11. http://pib. nic. in/newsite/PrintRelease. aspx?relid=132799.

Himachal Pradesh specialities. 2012. Himachal cuisine. mapsofindia. com. Retrieved , 2012.

Hirschfelder G and Schönberger GU. 2005. Germany. Sauerkraut, beer and so much more. In: Culinary cultures of Europe. Identity, diversity and dialogue (Goldstein C and Merkle K, Eds.). Council of Europe Publishing, Verlagsgruppe Lübbe, Germany. 183-194.

Hoddinott J, Maluccio J B, Jere F, Rafael M R. 2008. Effect of a nutritional intervention during early childhood on economic productivity among guatemalan adults. Lancet. 371: 411-6. 10. 1016/S0140-6736(08)60205-6.

Hoddinott J, Rosegrant M and Torero M. 2012. hunger and malnutrition. assessment paper for the 3rd Copenhagen consensus, copenhagen consensus center. http://www. copenhagenconsensus. com/sites/default/ files/hungerandmalnutrition. pdf.

Holleman C and Conti V. 2019. Role of income inequality in shaping outcomes on food insecurity. FAO Agricultural Development Economics Working Papers 19-06. Rome, FAO.

Holzer P, Pabst MA and Lippe I Th. 1969. Intragastric capsaicin protects against aspirin-induced lesion formation and bleeding in the rat gastric mucosa. Gastroenterology. 96:14i-33.

Hooper L, Summerbell CD, Thompson R, et al. 2012. Reduced or modified dietary fat for preventing cardiovascular disease (Review). The cochrane database of systematic reviews 5(CD002137): 1-3.

Hossain F, Muthusamy V, Pandey N, Vishwakarma AK, Baveja A, Zunjare RU, et al. 2018. Marker-assisted introgression of opaque2 allele for rapid conversion of elite hybrids into quality protein maize. J Genet. 97:287–98.

House LR. 1985. A Guide to sorghum breeding. International crops research institute for the semiarid tropics, Patancheru, India. 206.

Hout W. 2015. Putting political economy to use in aid policies. A governance practitioner's notebook: alternative ideas and approaches © OECD 2015.

http://apps. who. int/iris/bitstream/10665/44626/1/9789241501798_eng. pdf

http://apps. who. int/iris/bitstream/10665/44664/1/9789241501767_eng. pdf. ,WHO. Guideline: neonatal vitamin A supplementation. 2011. p. 1–25.

http://blog. tourism. gov. my/banana-leaf-rice/

http://books. google. co. in/books?hl=en&lr=&id=5WY08iuJyawC&oi=fnd&pg=PA41&dq=cardamom+nausea&ots=1LrJ-tcWSz&sig=uPPJCuNhxy6_m3cB9vONagsjWes#v=onepage&q=cardamom%20nausea&f=false

http://books. google. co. in/books?id=WJP7tgjzE_oC&pg=PA64&lpg=PA64&dq=cardamom+hiccups&Source=bl&ots=NsgzAII4lj&sig=Q1VoRwLIwBaG1ZDqvj2oYWXIJ0w&hl=en&sa=X&ei=1LBFUN1qieSsB-HwgdgF&ved=0CC8Q6AEwAA#v=onepage&q=cardamom%20hiccups&f=false

http://marel. in/

http://mofpi. nic. in/sites/default/files/databank. pdf.

http://naturalingredient. org/wp/wp-content/uploads/cardamom-benefits-_-information. pdf.

http://nmj. umsha. ac. ir/browse. php?a_id=1280&sid=1&slc_lang=en

http://nopr. niscair. res. in/bitstream/123456789/19002/1/IJEB%2037%283%29%20238-242. pdf

http://pib. nic. in/newsite/PrintRelease. aspx?relid=16151214

http://snackfoodmachines. com/

http://socialprotection. org/connect/communities/working-group-food-security-and-nutrition-specific-inter-agency-social

http://www. bajajmachines. com/

http://www. bucherindustries. com/en/about-us/locations

http://www. bucherindustries. com/en/divisions/bucher-specials

http://www. buhlergroup. com/global/en/home. htm#. WdB0sWiCzIU

http://www. economode. in/

http://www. esfoodtechnologies. com ; http://www. esfoodtech. com/

http://www. eurekaselect. com/146591/article

http://www. fao. org/3/a-i7212e. pdf

http://www. fao. org/economic/ptop/home/about/en/

http://www. frickweb. com/meat. aspx

http://www. gea. com/en/news/corporate-news/2017/gea-to-build-asia-largest- milk-productionfacility-in-india. jsp 7;

http://www. heatandcontrol. com/contact. asp

http://www. imedpub. com/articles/herbal-medicines-possible-risks-andbenefits. pdf

http://www. indianspices. com/spice-catalog/cardamom-small-1

http://www. indopol. com ; https://www. zaubacorp. com

http://www. irl. co. in/.

http://www. jarvis. co. in
http://www. jarvisproducts. com/ ; https://www. jarvis. co. in/
http://www. jbtfoodtech. com/
http://www. jmb. or. kr/journal/viewJournal. html?doi=10. 4014/jmb. 1605. 05056
http://www. mazmach. com/index. html
http://www. mofpi. nic. in/sites/default/files/important_notice-sampada-
http://www. ncbi. nlm. nih. gov/pmc/articles/PMC3818611/. Accessed August 27, 2014.
http://www. rediff. com/money/2007/may/30farm. htm
http://www. sanjivanindustries. com
http://www. satakeindia. com/
http://www. sifterinternational. com/index. php
http://www. sspindia. com/about-us. html
http://www. tetrapak. com/in
http://www. truefood. eu/
http://www. who. int/maternal_child_adolescent/documents/levels_trends_child_mortality_2015/en/
http://www. who. int/mediacentre/factsheets/malnutrition/en/
http://www. zaubacorp. com
http://www0. sun. ac. za/vivus
https://core. ac. uk/download/pdf/60543480. pdf
https://food52. com/blog/16041-for-middle-eastern-style-coffee-add-cardamom
https://ndb. nal. usda. gov/ndb/foods/show/02006
https://onlinelibrary. wiley. com/doi/abs/10. 1002/jsfa. 8165
https://onlinelibrary. wiley. com/doi/abs/10. 1002/jsfa. 8414
https://www. anko. com. tw/en/country/IN. html
https://www. atlaspacific. com/
https://www. ft. com/content/584ed992-4216-11e3-9d3c-00144feabdc0,
https://www. hrsasia. co. in/
https://www. ibef. org/industry/indian-food-industry. aspx
https://www. jarvis. co. in/
https://www. mdpi. com/2079-6382/8/1/8
https://www. meyn. com/
https://www. ncbi. nlm. nih. gov/pmc/articles/PMC3170500/pdf/nihms307509. pdf
https://www. ncbi. nlm. nih. gov/pmc/articles/PMC3353705/
https://www. ncbi. nlm. nih. gov/pmc/articles/PMC3975278/pdf/1749-8546-9-10. pdf
https://www. ncbi. nlm. nih. gov/pmc/articles/PMC4661898/pdf/ijms-16-26040. pdf
https://www. ncbi. nlm. nih. gov/pmc/articles/PMC4827078/pdf/EXCLI-12-1043. pdf
https://www. ncbi. nlm. nih. gov/pmc/articles/PMC5048087/pdf/fmicb-07-01580. pdf
https://www. ncbi. nlm. nih. gov/pmc/articles/PMC5775110/pdf/main. pdf
https://www. ncbi. nlm. nih. gov/pmc/articles/PMC6177529/
https://www. ncbi. nlm. nih. gov/pmc/articles/PMC6278479/pdf/molecules-23-02818. pdf
https://www. ncbi. nlm. nih. gov/pubmed/27246748
https://www. ncbi. nlm. nih. gov/pubmed/28603121
https://www. researchgate. net/publication/312151655_The_effect_of_Elettaria_cardamomum_extract_on_anxiety-like_behavior_in_a_rat_model_of_post-traumatic_stress_disorder
https://www. researchgate. net/publication/321573429_Processing_and_Medicinal_Uses_of_Cardamom_and_Ginger-A_Review
https://www. researchgate. net/publication/7655083_Cardamom_extract_as_inhibitor_of_human_platelet_aggregation.
https://www. sciencedirect. com/science/article/abs/pii/S1043661896900675?via%3Dihub.

https://www. sciencedirect. com/science/article/pii/S0753332216329535
https://www. sciencepubco. com/index. php/ijbas/article/view/5435
https://www. unicef. org/role. html

Hu FB, van Dam RM and Liu S. 2001. Diet and risk of Type II diabetes: the role of types of fat and carbohydrate. Diabetologia. 44(7):805-817.

Hummelbrunner R and Jones H. 2013. A guide for planning and strategy development in the face of complexity. Background Note. March 2013. London overseas development institute. International Food Policy Research Institute (IFPRI). 2014. Global Nutrition Report 2014: Actions and accountability to accelerate the world's progress on nutrition. Washington DC.

Igwe OU and Akabuike HC. 2016. Free radical scavenging activity, phytochemistry and antimicrobial properties of Tetrapleura tetraptera seeds. Intern Res J Chem Chemi Sci. 3:037–042.

Ilic N, Schmidt B M, PoulevA and RaskinI. 2010. Toxicological evaluation of grains of paradise (Aframomum melegueta). J Ethnopharmacol. 127: 10–17. doi:10. 1016/j. jep. 2009. 10. 031

Ilic NM, Dey M, PoulevAA, LogendraS, Kuhn PE and RaskinI. 2014. Anti-inflammatory activity of grains of paradise (Aframomum melegueta Schum) extract. JAgric Fd Chem. 62: 10452–10457. doi:10. 1021/jf5026086

Imbrasiene B. 2005. Lithuania. Rituals and feasts. In: Culinary cultures of Europe. Identity, diversity and dialogue (C Goldstein and K Merkle, eds.). Council of Europe Publishing, Verlagsgruppe Lübbe, Germany. 265- 279.

India, (report prepared by) Planning Commission, Government of Kerala development report. New Delhi. 2008. Academic Foundation. ISBN 81-7188-594-2.

India's bountiful breakfast options. The Times of India. 9 May 2010. Retrieved, 2012.

Indian Cuisine:Puducherry. 2011. India style cuisine.

Indian Food and the West New Emissary. 2010. The Diplomat. Retrieved , 2012.

Indian Street Food in Delhi. A tour of authentic Indian street food in Delhi. Foodtourindelhi. com. Retrieved, 2013.

Indiastat. 2018. Socio-economic statistical information about India. Accessed on October 12. https://www. indiastat. com/

Intakes for Calcium and Vitamin D. 2010. Washington, DC:National Academy Press.

International Food Policy Research Institute. 2018. Global food policy report. Washington, DC:Internl Fd Pol Res Ins. https://doi. org/10. 2499/9780896292970.

Ishara J S, Daniel K, Glaston B and Ariel N M G. 2018. Nutritional and physical attributes of maize-mushroom complementary porridges as influenced by mushroom species and ratio. Amer J Fd Nutr. 6. 17-27. 10. 12691/ajfn-6-1-4.

Issac Y A. 2012. Characterization and HPLC quantification of piperine in various parts of Piper guineenese. Department of Pharmarceutical Chemistry, Kwame Ukruma University of Science and Thechnology, Kumasi, Ghana.

Iweala E E J, Liu F F, Cheng R R, LiY, Ononhinmin C A, Zhand Y J. 2015. Anticancer and free radical scavenging acivity of some Nigerian food plants in vitro. Internl J Can Res. 11: 41–51.

Iwu M M. 2014. Handbook of African medicinal plants. Francis and Taylor Group, Boca Raton, FL. 279–333.

Iyer A, Panchal S, Poudyal H and Brown L. 2009. Potential health benefits of Indian spices in the symptoms of the metabolic syndrome:a review. Ind J Biochem Biophy. 46: 467–481.

Jackson RM, Ishara D N, Glaston S and Kenji M. 2018. Publisher: LAP Lambert Academic Publishing. Edible mushrooms: new food fortification approach toward food security. ISBN: 978-613-8-38655-1.

Jacobs and Fraikin 2005. Synthesis Report No 6. Traditional Foods in Europe Dr. Elisabeth Weichselbaum and Bridget Benelam British Nutrition Foundation Dr. Helena Soares Costa. National Institute of Health (INSA), Portugal.

Jakobsen M, O'Reilly E, Heitmann B, et al. 2009. Major types of dietary fat and risk of coronary heart disease: a pooled analysis of 11 cohort studies. The Amer J Clin Nutr. 89(5): 1425-1432.

Jenna F. 2018. What are the benefits of oatmeal?Medical News today. Rev. Natalie Olsen).

John S, Chi-Tang Ho, Fereidoon S. 2010. Functional Foods of the East. CRC Press. 64–. ISBN 978-1-4200-7192-4. Retrieved, 2012.

Johns T and Eyzaguirre PB. 2006. Linking biodiversity, diet and health in policy and practice. Proc Nutr Soc. 65: 182–189.

Johns T, Powell B, Maundu P et al. 2013. Agricultural biodiversity as a link between traditional food systems and contemporary development, social integrity and ecological health. J Sci Food Agric. 93: 3433–3442.

Johnston D, Stevano S, Malapit H, Hull E and Kadiyala S. 2015. Agriculture, gendered time use and nutritional outcomes: A Systematic Review. (IFPRI discussion paper No. 01456). http://ebrary. ifpri. org/utils/getfile/collection/p15738coll2/id/129389/filename/129600. pdf

Jónsson OD. 2005. Iceland. The creative fight for survival. In: Culinary cultures of Europe. Identity, diversity and dialogue (Goldstein C and Merkle K, Eds.). Council of Europe Publishing, Verlagsgruppe Lübbe, Germany. 217-225.

Juárez-López C, Klünder-Klünder M, Madrigal-Azcárate A, et al. 2013. Omega-3 polyunsaturated fatty acids reduce insulin resistance and triglycerides in obese children and adolescents. Pediatric Diabetes 14(5):377-383.

Juma C. 2016. Technology will drive scientific progress in Africa not the other way round. Quartz Africa. https://qz. com/africa/.

Kabeer N. 1999. ReSources, agency, achievements:reflections on the measurement of women's empowerment. Dev. Change. 30: 435-464.

Kabiru AY, Ibikunle GF, InnalegwuDA, Bola BM and Madaki F M. 2016. *In vivo* antiplasmodial and analgesic effect of crude ethanol extract of Piper guineense leaf extract in Albino mice. Scientifica (Cairo)86:87–93.

Kaefer C M and Milner J A. 2011. Herbs and spices in cancer prevention and treatment. Chapter 17. In:Benzie, I F F, Wachtel-Galor, S. (Eds.) Herbal Medicine: Biomolecular and Clinical Aspects. CRS Press/Taylor and Francis, Boca Raton, FL.

Kalupahana N, Claycombe K and Moustaid-Moussa N. 2011. (n-3) Fatty acids alleviate adipose tissue inflammation and insulin resistance: mechanistic insights. Advances in Nutrition : An International Review Journal 2(4): 304-316.

Kamtchouing P, Mbongue G Y, DimoT, WatchoP, Jatsa H B and Sokeng S D. 2002. Effects of Aframomum melegueta and Piper guineense on sexual behaviour of male rats. Behavl Pharmacol. 13: 243–247.

Kang JY, Teng CH, Wee A and Chen FC. 1995. Effect of capsaicin and chilli on ethanol induced gastric mucosal injury in the rat. Gut. 36(5):664–669. doi:10. 1136/gut. 36. 5. 664.

Kashmiri food. Kashmiri cuisine. food-india. com. Retrieved, 2012.

Kashmiri Wazwan. Metroplus. Chennai, India: thehindu. com. 11 April 2012. Retrieved, 2012.

Kathleen R and O'Brien R. 2018. Healthwise Staff. Med Rev.

Kaur N, Chugh V and Gupta AK. 2014. Essential fatty acids as functional components of foods-a review. J Fd Sci Technol. 51(10):2289–2303. doi:10. 1007/s13197-012-0677-0

Keatinge J D H, Yang R Y, Hughes J dA, Easdown W J and Holmer R. 2011. The importance of vegetables in ensuring both food and nutritional security in attainment of the Millennium Development Goals. Fd Sec. 13: 4.

Keith Pearson. 2017. UNICEF. UNICEF and Pearson Launch Educational Partnership for Children Equipping marginalised youth in Myanmar and Vietnam with 21st century and employability skills.

Keller G. 2004. African nightshade, eggplant, spider flower—production and consumption of traditional vegetables in Tanzania from the farmers point of view. MSc. Thesis, Universität, Göttingen, Germany.

Khan BA, Abraham A and Leelamma S. 1995. Hypoglycemic action of Murraya koenigii (curry leaf) and Brassica juncea (mustard): mechanism of action. Ind J Biochem Biophy. 32(2):106-108.

Khush GS. 2015. IRRI and the green revolution in India. http://irri. org/rice-today/india/ revisiting-50-years-of-indian-rice-research/irri-and-the-green-revolution-in-india?tmpl=component&print=1&page=.

Kleijnen J, Knipschild P and ter Riet G. 1989. Garlic, onions and cardiovascular risk factors. A review of the evidence from human experiments with emphasis on commercially available preparations. Br J Clin Pharmacol. 28(5):535-44.

Klin-Kabari B B, Barimalaa I S and Achenwu S C. 2011. Effect of three indigenous spices on the chemical stability of smoked dried catfish (Clarias lazera) during storage. AfrJ Agri Nutri Develop. 11: 5–9.

Kochhar KP. 1996. An Experimental Study on some Physiological Effects of Dietary Spices. Thesis submitted to AIIMS for the Degree of Doctor of Philosophy.

Kochhar K P. 2008. Dietary spices in health and diseases (II). Ind J Physiol Pharmacol. 52: 327–354.

Kraemer K and Zimmermann MB. 2007. Nutritional Anemia. Basel, Switzerland: Sight and Life Press. ISBN 3-906412-33-4.

Krishna Gopal Dubey. 2011. The Indian Cuisine. PHI Learning Pvt. Ltd. 86–232. ISBN 978-81-203-4170-8. Retrieved, 2012.

Krishna P B and Sukhdev S C. 1977. Caste, Tribes and Culture of India:North-Eastern India. Ess Publications. 244. Retrieved, 2012.

Krishnaswamy. 2008. Traditional Indian spices and their health significance. Asia Pac J Clin Nutr. 17(Suppl 1):265-8.

Kristjansson E A, Kristjansson B, Petticrew M, MacDonald B, Krasevec J, Janzen L, Greenhalgh T, Wells GA, MacGowan J, Farmer A P, Shea B, Mayhew A, Tugwell P and Welch V. 2006. School feeding for improving the physical and psychosocial health of disadvantaged elementary school children. Campbell System Rev. 14: 1–135.

Krzysztofek K. 2005. Poland. Cuisine, culture and variety on the Wisla river. In: Culinary cultures of Europe. Identity, diversity and dialogue (C. Goldstein and K. Merkle, Eds.). Council of Europe Publishing, Verlagsgruppe Lübbe, Germany. 333-346.

Kuete V. et al. 2011. Cytotoxicity of some Cameroonian spices and selected medicinal plant extracts. J Ethnopharmacol. 134:803–812.

Kumar Suresh Singh. 1994. People of India:Daman and Diu. Popular Prakashan. 8–9. ISBN 978-81-7154-761-6. Retrieved, 2012.

Kumar V, Rani A, Shuaib M and Mittal P. 2018. Comparative assessment of trypsin inhibitor vis-à-vis Kunitz trypsin inhibitor and bowman-birk inhibitor activities in soybean. Fd Anal Method. 11:2431–7.

Lachat C, Otchere S, Roberfroid D, et al. 2013. Diet and physical activity for the prevention of noncommunicable diseases in low and middle income countries: a systematic policy review. PLoS Med. 10(6):e1001465. doi:10. 1371/journal. pmed. 1001465

Lachu M and Snortum M. 2005. Ajanta: Regional Feasts of India. Gibbs Smith. 59. ISBN 978-1-58685-777-6. Retrieved, 2012.

Lance Richardson. Traveller. Best traditional USA dishes:Top ten must try American dishes. www. traveller. com. au

Laura K. 2009. The Silk Road Gourmet:Western and Southern Asia. iUniverse. 298. ISBN 978-1-4401-4305-2. Retrieved, 2012.

Leach RM and Harris ED. 1997. Manganese. In: Handbook of nutritionally essential minerals. Eds. O'Dell BL, Sunde RA. New York: Marcel Dekker, Inc:335-355.

Lefevre M, Champagne C, Tulley R, et al. 2005. Individual variability in cardiovascular disease risk factor responses to low-fat and low-saturated-fat diets in men: body mass index, adiposity, and insulin resistance predict changes in LDL cholesterol.

Lemnis M and Vitry H. 1981. Old Polish traditions in the kitchen and at the table. Interpress Publishers, Warsaw.

Leveraging Small and Medium Enterprises to improve nutrition. 2018. Published by the Food and Agriculture Organization of the United Nations and the Global Alliance for Improved Nutrition, Rome.

Lidell ME, Betz MJ, Leinhard OD, et al. 2013. Evidence for two types of brown adipose tissue in humans. Nat Med. 19(5):631-634.

Lindsay B and Amelia T. 2008. Rajasthan, Delhi and Agra. Lonely Planet. 69. ISBN 978-1-74104-690-8. Retrieved, 2012.

Linhart Y B, Gauthier P, Keefover-Ring K and ThonpsonJ D. 2015. Variable phytotoxic effects of Thymus vulgaris (Lamiaceae) terpenes on associated species. Intern J Pl Sci. 176: 20–30.

Lisa MH. 2003. Exotic Appetites:Ruminations of a Food Adventurer. Routledge. 34. ISBN 978-0-415-94385-7. Retrieved, 2012.

Little TJ, Gopinath A, Patel E, et al. 2010. Gastric emptying of hexose sugars: role of osmolality, molecular structure and the CCK receptor. Neurogastroenterol Motil. 22:1183-1190, e314.

Liu YQ, Davis CR, Schmaelzle ST, Rocheford T, Cook ME, Tanumihardjo SA, et al. 2012. β-cryptoxanthin biofortified maize (Zea mays) increases β-cryptoxanthin concentration and enhances the color of chicken egg yolk. Poult Sci. 91:432–8.

Liyanage C and Hettiarachchi M. 2011. Food fortification. Ceylon Med J. 56: 124-127.

Lonely Planet blue list: the best in travel. 2007. Lonely Planet. 2006. Retrieved, 2012.

Lordan R, Tsoupras A, Mitra B, Zabetakis I. 2018. Dairy Fats and Cardiovascular Disease:Do We Really Need to be Concerned?. Foods. 7(3):29. doi:10. 3390/foods7030029

Low JW, Arimond M, Osman N, Cunguara B, Zano F, Tschirley D, et al. 2007. A food-based approach introducing orange-fleshed sweet potatoes increased Vitamin A intake and serum retinol concentrations in young children in rural Mozambique. J Nutr. 137:1320–7.

Madkor HR, Mansour SW and Ramadan G. 2011. Modulatory effects of garlic, ginger, turmeric and their mixture on hyperglycaemia, dyslipidaemia and oxidative stress in streptozotocin-nicotinamide diabetic rats. Br J Nutr. 105(8):1210-7. Epub 2010.

Mahady GB, Pendland SL, Stoia A, Hamill FA, Fabricant D, Dietz BM and Chadwick LR. 2005. In vitro susceptibility of Helicobacter pylori to botanical extracts used traditionally for the treatment of gastrointestinal disorders. Phytother Res. 19(11):988-91.

Maharishi Mahesh Yogi on the Bhagavad Gita Translation and Commentary, Arkana, 1990. 236.

Majumdar B. 2013. Cooking on the run (1st ed.). Delhi: Harper Collins India. 192.

Malapit HJL and Quisumbing AR. 2015a. What dimensions of women's empowerment in agriculture matter for nutrition in Ghana?Fd Policy. 52: 54-63. https://doi. org/10. 1016/j. foodpol. 2015. 02. 003

Malapit L, HazelJean S, Kadiyala AR, Quisumbing K, Cunningham P Tyagi. 2015b. Women's empowerment mitigates the negative effects of low production diversity on maternal and child nutrition in Nepal. JDev. Stud. 51 (8):1097-1123. https://doi. org/10. 1080/00220388. 2015. 1018904

Man E H and Ellis A J. 1932. The Aboriginal Inhabitants of the Andaman Islands. Mittal Publications.

Mann A. 2011. Biopotency role of culinary spices and herbs and their chemical constituents in health and commonly used spices in Nigerian dishes and snacks. Afri J Fd Sci. 5:111–124.

Marmot M B, Ellen G P. 2013. The Role of Social Determinants in Tackling Health Objectives in a Context of Economic Crisis. Public Health Rev. 35. 10. 1007/BF03391694.

Marotta RB, Floch MH. 1991. Diet and nutrition in ulcer disease. Med Clin North Am. 75(4):967-79.

Martin H, Sheema M and Richard D. 2001. World Food India. Lonely Planet. 176. ISBN 978-1-86450-328-9. Retrieved, 2012.

Martins. 2013. http://www. martinslibrary. com (accessed 20 December 2016).

Martorell R. 2017. Improved nutrition in the first 1000 days and adult human capital and health. AmJHumBiol. 29(2):10. 1002/ajhb. 22952. doi:10. 1002/ajhb. 22952

Masayi N and Netondo G. 2012. Effects of Sugarcane Farming on Diversity of Vegetable Crops in Mumias Division, Western Kenya. Internl J Biodiver Conserv. 4 (13): 515-24.

Masika R. 2002. Editorial:Gender and Climate Change. Gender and Development. 10 (2).

Mayo-Wilson E, Imdad A, Herzer K, Yakoob M, Bhutta Z. 2011. Vitamin A supplements for preventing mortality, illness and blindness in children aged under 5:systematic review and meta-analysis. BrMedJ. 343:d5094. http://www. bmj. com/content/343/bmj. d5094.

Mba M A. 1994. Effect of dietary intake of Piper guineense on growth and indices of fitness in Rattus rattus. isc. Innoa. 4: 383–388.

Mbongue F G Y, Kamtchouing P, Essama O J L, YewahP M, Dimo T and Lontsi, D. 2005. Effect of the aqueous etract of dry fruits of Piper guineense on the reproductive function of adult male rats. Ind J Pharmacol. 37:30–32.

McCarty MF, DiNicolantonio JJ, O'Keefe JH. 2015. Capsaicin may have important potential for promoting vascular and metabolic health. Open Heart. 2(1):e000262. doi:10. 1136/openhrt-2015-000262.

McKingney JI. 1983. Intervention for the prevention of vitamin A deficiency: a summary of experiences. In: Underwood BA, ed. Nutrition intervention strategies in national development. New York: Academic Press. 363-84.

Megan Metropulos. 2018. Cloves:Nutritional facts and health benefits. Medically reviewed by Natalie Butler.

Mensink RP, Zock PL, Kester AD, et al. 2003. Effects of dietary fatty acids and carbohydrates on the ratio of serum total to HDL cholesterol and on serum lipids and apolipoproteins: a meta-analysis of 60 controlled trials. The Am J Clin Nutr. 77(5):1146-55.

Merah O. 2015. Genetic variability in glucosinolates in seed of Brassica Juncea:Interest in mustard condiment. JChem. Article ID 606142, 6 pages. https://doi. org/10. 1155/2015/606142.

Meredith C. and Kaiser RN. 2019. How food affects blood sugar. Clinical review. Reviewed 01/03/2019.

Messina MJ. 1989. Legumes and soybeans:overview of their nutritional profiles and health effects. Am J Clin Nutr. American Society for Clinical Nutrition, Inc 49: 725-737.

Micha R and Mozaffarian D. 2010. Saturated fat and cardiometabolic risk factors, coronary heart disease, stroke, and diabetes: a fresh look at the evidence. Lipids - Springer 45(10): 893-905.

Micha R, Khatibzadeh S, Shi P, et al. 2014. Global, regional and national consumption levels of dietary fats and oils in 1990 and 2010: a systematic analysis including 266 country-specific nutrition surveys. BMJ:Briti Med J. 348.

Moehn S, Pencharz PB and Ball RO. 2012. Lessons learned regarding symptoms of tryptophan deficiency and excess from animal requirement studies. J Nutr. 142:2231S–2235S.

Mohammed A, Koorbanally N A and Islam M S. 2015. Ethyl acetate fraction of Aframomum melegueta fruit ameliorates pancreatic β-cell dysfunction and major diabetes-related parameters in a type 2 diabetes model of rats. J Ethnopharmacol. 175:518–27.

MohantyA. 2017. Rasagola:The Ritual Offering of Odisha (PDF). Odisha Review: 76–94. ISSN 0970-8669. Archived from the original (PDF) on 2 November 2017. Retrieved, 2017.

Mohebi-Nejad A and Bikdeli B. 2014. Omega-3 supplements and cardiovascular diseases. Tanaffos. 13(1):6–14.

Mojekwu T O, YamaO E, OjokukuS A, Oyebadejo S A. 2011. Hypoglyceamic effects of aqueous extract of Aframomum melegueta leaf on alloxan-induced diabetic male albino rats. Pacif J Med Sci. 8:28–35.

Mozaffarian D, Aro A and Willett WC. 2009. Health effects of trans-fatty acids: experimental and observational evidence. Eur J Clin Nutr. 63 (Suppl 2): S5-21.

Mozaffarian D, Lemaitre R, King I, et al. 2013. Plasma phospholipid long-chain ω-3 fatty acids and total and cause-specific mortality in older adults: a cohort study. Annals Intern Med. 158(7): 515-25.

Mughlai cuisine tops popularity charts in capital. 2008. Telegraph India. 5 July Retrieved, December 2013.

Muller HG. 1969. Application of the statistical theory of the rubber elasticity to gluten and dough. Ceareal Chem. 46: 443.

Murdoch Books Pvt. Limited. 2010. Murdoch Books Test Kitchen. India. Murdoch Books. 9. ISBN 978-1-74196-438-7. Retrieved, 2012.

Musinguzi E, Kikafunda J K and Kiremire B T. 2006. Utilization of indigenous food plants in uganda: a case study of South-western Uganda. Afri J Fd Agric Nutr Develop. 6 (2): 1-21.

Muthusamy V, Hossain F, Thirunavukkarasu N, Choudhary M, Saha S, Bhat JS, et al. 2014. Development of β-carotene rich maize hybrids through marker-assisted introgression of β-carotene hydroxylase allele. PLoS One. 9:e113583.

Mwaseba DJB and Kaarhus R. 2015. How do intra-household gender relations affect child nutrition?Findings from two rural districts in Tanzania. ForumDevStud. 42 (2): 289-309. https://doi. org/10. 1080/08039410. 2015. 1020337

Natalie B RD LD. 2018. Health benefits and nutritional value of spinach. Written by Megan.

Nathan CS. 2013. Chettinad fare to tempt palate. The Star. Retrieved, 2014.

National Academy Press, 2010. Promoting cardiovascular health in the developing world: a critical challenge to achieve global health. Kelly BB and Fuster V.

National family health survey-4, 2015-16. NFHS-4 fact sheets for key indicators based on final data Mumbai: IIPS. 2017. International Institution for Popul Sci.

National Health and Medical Research Council, Department of Health and Ageing. 2013. Eat for health australian dietary guidelines providing the scientific evidence for healthier Australian diets. Canberra, Australia.

National Institute of Health, Office of Dietary Supplements). New PubMed® subset - "Dietary Supplements" (U. S. Government content and in the Public Domain).

National Institutes of Health; Office of Dietary Supplements. Jost Chemical Co. The importance of minerals in the long term health of humans. Phillipe H Merrell.

Nedergaard J, Bengtsson T and Cannon B. 2007. Unexpected evidence for active brown adipose tissue in adult humans. Amer J Physiol Endocrin Metab. 293(2):E444-E452.

Neeraja CN, Ravindra BV, Ram S, Hossain F, Hariprasanna K, Rajpurohit BS, et al. 2017. Biofortification in cereals-Progress and prospects. Curr Sci. 13:1050–7.

Newman D J and Cragg G M. 2012. Natural products as Sources of new drugs over the 30 years from 1981 to 2010. J Natl Prod. 75:311–335.

Ngassoum M, Jirovetz L and Buchbauer G. 2001. SPME/GC/MS analysis of headspace aroma compounds of the Cameroonian fruit Tetrapleura tetraptera (Thonn) Taub. Eur Fd Res Technol. 213: 18–29.

Nickel E H. 1995. Publication approved by the Commission on New Minerals and Minerals Names of the International Mineralogical Association).

NIH Consensus Statement: 1994. 12:1-31.

Nordic Council of Ministers. 2012. Nordic Nutrition Recommendations. Part 1. 5th Edition. Copenhagen, Denmark.

Norman M P. 1980. Poison-Damsels. Ayer Publishing. 253. ISBN 978-0-405-13336-7. Retrieved, 2012.

Nuss ET and Tanumihardjo SA. 2011. Quality protein maize for Africa: Closing the protein inadequacy gap in vulnerable populations. Adv Nutr. 2:217–24.

Nwakwo CS, Ebenezer I A, Ikpeama AI and Asuzu FO. 2014. The nutritional and antinutritional values of two culinary herbs – uziza leaf (Piper guineese) and scent leaf (Ocimum gratissimum) popularly used in Nigeria. Internl J Sci Eng Res. 5: 1160–1163.

O'Mahony, Rachel and Al-Khtheeri, Huda and Weerasekera, Deepaka and Fernando, Neluka and Vaira, Dino and Holton, John and Christelle and Basset. 2006. Bactericidal and Anti-adhesive properties of culinary and medicinal plants against Helicobacter pylori. World J Gastroenterol. WJG. 11: 7499-507.

O'Neill J, Brock C, Olesen AE, Andresen T, Nilsson M and Dickenson AH. 2012. Unravelling the mystery of capsaicin:a tool to understand and treat pain. Pharmacol Rev. 64(4):939–971. doi:10. 1124/pr. 112. 006163

OECD, Organisation for Economic Co-operation and Development. 2008. Annual report on sustainable development work in the OECD.

Ogbera A O and Ekpebegh C. 2014. Diabetes mellitus in Nigeria: the past, present and future. World J Diab. 5: 905–911. doi:10. 4239/wjd. v5. i6. 905

Ogbunugafor H A. et al. 2011. Physico-chemical and antioxidant properties of Moringa oleifera seed oil. Pakistan J Nutri. 10: 409–414.

Oguoma OI, Ezeifeka G O, Adeleye S A, Oranusi S and Amadi E S. 2015. Antimicrobial activity, proximate and amino acid analysis of T. tetraptera. Nigerian J Microbiol. 27:2709–2718.

Ojewole J A. 2005. Analgesic and anticonvulsant properties of Tetrapleura tetraptera (taub) (fabaceae) fruit aqueous extract in mice. Phytotherapy Res. PTR19:1023–1029.

Ojewole JA and Adewunmi CO. 2004. Anti-inflammatory and hypoglycaemic effects of Tetrapleura tetraptera (taub) [fabaceae] fruit aqueous extract in rats. J Ethnopharma. 95: 177–182.

Okokon J E, Udokpoh A E and Antia B S. 2007. Antimalaria activity of ethanolic extract of Tetrapleura tetraptera fruit. J Ethnopharmacol. 111: 537–540.

Olawuyi E B, Osikabor B and AderoumuAF. 2013. Effects of Piper guineense (Schumach) seeds extract on the control of insect pest of Amaranthus viridis (Linn). J Forestry Res Manage. 10: 83–92.

Olumakaiye M F and Ajayi AO. 2006. Women's empowerment for household food security: the place of education. J Hum Ecol. 19 (1):51-55.

Omotayo O A, Adepoju O, ThomasK and Oluremi O. 2013. Evaluation of micronutrient potentials of seven commonly consumed indigenous spices from Nigeria. Amer J Fd Nutr. 3: 122–126.

Organic Facts on Internet: Arthritis: The health benefits of copper related to its anti-inflammatory actions that assist in reducing the symptoms of arthritis. Meenakshi Nagdeve. Updated - January 29, 2020.

Orwa C, Mutua A, Kindt R, JamnadassR andAnthonyS. 2009. Agroforestree Database: a tree reference and selection guide version. World Agroforest Center, Kenya.

oyhus E-M. 2005. Denmark, Nation-building and cuisine. In: Culinary cultures of Europe. Identity, diversity and dialogue. Eds. Goldstein C and Merkle K. Council of Europe Publishing, Verlagsgruppe Lübbe, Germany. 129-137.

Padmanabh S Jaini. 2000. Collected papers on Jaina Studies. Motilal Banarsidass. ISBN 978-81-208-1691-6: 3–14.

Padmanabh S Jaini. 2001. Collected papers on Buddhist Studies, Motilal Banarsidass, ISBN 978-81-208-1776-0: 57–77.

Pal D and Verma P. 2013. Flavonoids: a powerful and abundant Source of antioxidants. Intern J Pharm Pharmaceut Sci. 5: 97–106.

Palmer AC, Craft NE, Schulze KJ, Barffour M, Chileshe J, Siamusantu W, et al. 2018. Impact of biofortified maize consumption on serum carotenoid concentrations in Zambian children. Eur J Clin Nutr. 72:301–3.

Pamela G K, Kathryn P S and Marcia N. 2011. Food and Culture. Cengage Learning. ISBN 978-0-538-73497-4: 449. Retrieved, 2012.

Pandey A. 2018. Tribal Cooking of Dadra and Nagar Haveli. ISSN 2348-3156 (Print) Intern J Socl Sci Human Res. ISSN 2348-3164(online). 6(2):670-675. www. researchpublish. com

Pappas LS. 2008a. Ethnic Cuisine: Belgium. Online: http://www. sallys-place. com/food/cuisines/belgium. htm.

Pappas LS. 2008b. Ethnic Cuisine: Spain. Online http://www. sallys-place. com/food/cuisines/spain. htm.

Paramahansa Yogananda. 1973. Autobiography Of A Yogi. Self Realization Fellowship. 22.

Paroda RS and Joshi PK(Eds.). 2017. Proceedings of the national conference on sustainable development goals: India's preparedness and the role of agriculture. New Delhi, India: Indian Council of Agricultural Research (ICAR), Trust for Advancement of Agricultural Sciences (TAAS) and International Food Policy Research Institute (IFPRI):48.

Pat Chapman. 2009. India: Food and Cooking:The Ultimate Book on Indian Cuisine:38. New Holland Publishers. ISBN 978-1-84537-619-2. Retrieved , 2012.

Patel R C. 2012. Food Sovereignty: Power, Gender and the Right to Food. PLOS Medicine 9(6): e1001223. DOI:10. 1371/journal. pmed. 1001223

Paul J, Khanna H, Kleidon J, Hoang P, et al. 2017. Golden bananas in the field: elevated fruit pro-vitamin A from the expression of a single banana transgene. Plant Biotechnol J. 15:520–32. doi:10. 1111/pbi. 12650.

Peter FD. 1985. Innovative and Entrepreneurship Practice and Principles. Harper and Row, Publishers, Inc.

Peterman A, Behrman J and Quisumbing A. 2010. A Review of Empirical evidence on gender differences in nonland agricultural inputs, technology, and services in developing countries. Washington DC: IFPRI.

Pfeiffer WH and McClafferty B. 2007. HarvestPlus: Breeding crops for better nutrition. Crop Sci. 47:S88–105.

Pinstrup-Andersen P and Watson II DD. 2011. Food policy for developing countries: The role of government in global, national, and local food systems. Cornell University Press. REACH. 2013. Country Facilitator Manual: 2nd edition. Rome, World Food Programme.

Plant Resources of Tropical Africa (PROTA). 2005. Comparative Data on 275 Vegetables. In Vegetables of Tropical Africa:Conclusions and Recommendations based on PROTA 2: Vegetables, edited by Bosch CH, Borus D J and Siemonsma J S. PROTA Foundation, Wageningen, The Netherlands.

Popay S E, Hernández M, Johnston H, Mathieson J and Rispel L. 2008. Understanding and Tackling social exclusion. Final Report to the WHO commission on social determinants of health. From the Social Exclusion Knowledge Network. Geneva, Switzerland, WHO. https://www. who. int/social_determinants/knowledge_networks/final_reports/sekn_final%20 report_042008. pdf?ua=1).

Potrykus I. 2014. From the concept of totipotency to biofortified cereals. Annu Rev Plant Biol. 66(1):1–22. http://www. annualreviews. org/doi/abs/10. 1146/annurev-arplant- 043014-114734.

Pran Nath Chopra. 1992. Encyclopaedia of India:Nagaland. Rima Publishing House. 63. Retrieved , 2012.

Prasad AS. 1996. Zinc deficiency in women, infants and children. J Am Coll Nutr. 15:113–20.

Press Trust of India. 2008. Goan fenny, Assam tea, Lucknow chikan to get GI registry soon. The Economic Times. New Delhi, India. Retrieved 2012.

Priti Chitnis Gress. 2008. Flavorful india: treasured recipes from a gujarati family. Hippocrene Books. 15. ISBN 978-0-7818-1207-8. Retrieved, 2012.

Priya Rampal 2018. An analysis of protein consumption in india through plant and animal Sources. Int. Nutr. Foun. Fd. and Nutr. Bull. 39(4).

Priya W and Carol SR. 2005. Food of India. Murdoch Books. p13. ISBN 978-1-74045-472-8. Retrieved, 2012.

Punjabi cuisine. 2012. Khana-Indian Recipes and Cuisines. Archived from the original.

Raboloko M. 2016. Determinants of urban household food security in Botswana. J Biomed Clin Sci. 1: 17–21.

Rajatananda Das Gupta. 1982. Art of mediaeval Assam. Cosmo. : 171. Retrieved, 2012.

Rasheed El and Salim, Rasheed and Abd, Widad and Ahmed, El and Adam, Mazahib and Al, Mohammed and Alsiddig, Taib and Hamed, Sara. 2017. Fortified sorghum as a potential for food security in rural areas by adaptation of technology and innovation in Sudan. J Fd Nutr Popul Health. 1:(1).

Ravi M. 2010. German froth Bitburger to debut in India. Retrieved, 2012.

Raybould HE and Tache Y. 1988. Capsaicin-sensitive afferent innervations of the stomach and stimulation of gastric acid secretion in the rat. Gastroenterol. 94: A371.

Reardon T, Barrett CB, Berdegué JA and Swinnen JFM. 2009. Agrifood industry transformation and small farmers in developing countries. World Develop. 37(11): 1717–27.

Reardon T, Timmer CP, Barrett CB and Berdegué J. 2003. The rise of supermarkets in Africa, Asia and Latin America. Amer J Agricul Econ. 85(5):1140-1146.

Renu Agrawal. 2016. Role of probiotics on improving human health. ISBN-13: 978-9385516719. Published by NIPA and CRC.

Renu Agrawal. and Sadali C. Basappa. 1994. Fermentation of paddy malt mash to ethanol by mixed cultures of Saccharomyces cerevisiae and Zymomonas mobilis ZM4 with penicillin G. J Ferment Bioeng. 77(2): 218-220.

Riaz MN. 1988. Soybeans as functional foods. Cereal Fds World. 44: 88-92.

Richter M. 1997. Who Milks the Cow? Gender-and-development in Livestock Farming. Schriftenreihe der GTZ. No. 261. Eschborn, Germany: German Agency for Technical Cooperation (GTZ). RNCOS. 2006. Indian Livestock Industry—An Industry Analysis. RNCOS Research. Noida, India: RNCOS.

Roberts R, Agre P, Alferof Z, et al. 2016. Support precision agriculture. Open online letter from 123 Nobel Leaureates, addressed to the leaders of Greenpeace, the United Nations and Governments around the world. http://supportprecisionagriculture. org/.

Rome Declaration on Nutrition, 2014. 39th Session of the FAO Conference: At the 39th session of the Conference of the FAO, participants will determine the policy and approve the budget of the Organization and make recommendations concerning questions relating to food and agriculture. dates: 6-13 June 2015 venue: FAO location: Rome, Italy e-mail: FAO-Conference@fao. org phone: +39 06 57051. fax: +39 06 570 53152. www:http:// www. fao. org/ unfao/govbodies/gsbhome/conference/en/

Rome Declaration on World Food Security and World Food Summit Plan of Action: World Food Summit 13-17 November, 1996. Rome. FAO. 2013. State of Food and Agriculture 2013: Food systems for better nutrition. Rome. FAO/WHO. 2014.

Ronteltap A T, Hans R and Lynn R F. 2007. Consumer acceptance of technology-based food innovations: Lessons for the future of nutrigenomics. Appetite. 49. 1-17. 10. 1016/j. appet. 2007. 02. 002.

Rooney LW and Mc Donough CM. 1987. Food quality and consumer acceptance of pearl millet. Proceedings of the International Pearl Millet Workshop Hyderabad, Inde, 7-11 April, : 43-61. Patancheru, Inde, ICRISAT.
Rosengarten Jr, Frederic. 1969. The Book of spices. Wynnewood, Pennsylvania, Livingston Publishing Company.
Rosgvist F, Iggman D, Kullberg J, et al. 2014. Overfeeding polyunsaturated and saturated fat causes distinct effects on liver and visceral fat accumulation in humans. Amer Diab Assoc. 63(7): 2356-68
Royal Mughal Ladies and Their Contributions. 2011. Retrieved 6 March 2015. 345 pages; Publisher: Gyan Publishing House. ISBN-10:8121207606 and ISBN-13: 978-8121207607.
Rubió L, Motilva M J and Romero M P. 2013. Recent advances in biologically active compounds in herbs and spices:a review of the most effective antioxidant and anti-inflammatory active principles. Crit Rev Fd Sci Nutr. 53: 943–953.
Ruzzene M. 2015. Beyond growth:problematic relationships between the financial crisis, care and public economies and alternative currencies. Internl J Commun Curren Res. 9: 81-93.
Ryan Raman. 2019. Healthline. 11 Impressive Health Benefits of Saffron.
S Salim EA. 2016. Organoleptic properties of fortified sudanese sorghum diets with wheat and soy.
Sae-Eaw A, Chompreeda P, Prinyawiwatkul W, Haruthaithanasan V, Suwonsichon T, Saidu JE and Xu Z, 2007. Acceptance and purchase intent of US consumers for nonwheat rice butter cakes. J Food Sci. 72 (2): S92-S97.
Saini N, Singh N, Kumar A, Vihan N, Yadav S, Vasudev S, et al. 2016. Development and validation of functional CAPS markers for the FAE genes in Brassica juncea and their use in marker-assisted selection. Breed Sci. 66:831–7.
Salam R and Das, Jai and Ali, Anum and Lassi, Zohra and Bhutta and Zulfiqar. 2014. Maternal undernutrition and intrauterine growth restriction. Expert Review of Obstetrics and Gynecology. 8. 10. 1586/17474108. 2013. 850857.
sales@fao. org©FAO, 2017.
Saltzman A, Birol E, Oparinde A, Andersson MS, Asare-Marfo D, Diressie MT, Gonzalez C, Lividini K, Moursi M, Zeller M and Ann N Y. 2017. Availability, production and consumption of crops biofortified by plant breeding: current evidence and future potential. Acad Sci. 1390(1):104-114.
Samuel A. 2013. Home is where the stomach is. New Straits Times. Retrieved, 2014.
Sancar FH. 2005. Turkey. The tastes of a splendid heritage. In: Culinary cultures of Europe. Identity, diversity and dialogue (C Goldstein and K Merkle, Eds). Council of Europe Publishing, Verlagsgruppe Lübbe, Germany. 443-457.
Sandoiu Ana. 2017. Study lists 33 foods proven to relieve rheumatoid arthritis. Med. News today. (9th November).
Sandra Bastin. Extension Food and Nutrition Specialist. Minerals in the diet. Educational programs of the Kentucky Cooperative Extension Service. Issued 03-96; FN-SSB. 028LG.
Sanjeev Dak Kapoor. Shin Delights. 1/e 295. 00. Popular Prakashan. p8. ISBN 978-81-7991-400-7. Retrieved, 2012.
Sanni MO and Sobamiwa AO. 1993. Processing and characteristics of soybean-fortified gari. World J Microbiol Biotechnol. 10: 268.
Sarda K. 2017. Chhaang:The Beer of the Himalayas. LiveStoryIndia. Retrieved, 2018.
Satyanarayana MN. 2006. Capsaicin and gastric ulcers. Crit Rev Fd Sci Nutr. 46(4):275-328.
Scalbert A and Williamson G. 2000. Dietary intake and bioavailability of polyphenols. The J Nutr. 130: 2073S–2085S.
Schneider A. and Ingram H. 1997. Policy design for democracy. University of Kansas Press.
Scrimshaw NS. 1984. Functional consequences of iron deficiency in human populations. J Nutr Sci Vitaminol. 30:47–63.

Seafood Curry(Malaysian Indian-Style)-Easy Delicious Recipes. rasamalaysia. com. Retrieved 23 March 2018.

Semba R. 2012. The vitamin A story: lifting the shadow of death. World Rev Nutr Diet. 104:207.

Shah N. 1998. Jainism: The World of Conquerors. Sussex Academic Press. p. 249. ISBN 978-1-898723-30-1. Retrieved, 2012.

Shankar SR, Yadav RK, Ray RB, Bijlani RL, Baveja T, Jauhar N, et al. 2005. Serum lipid response to introducing ghee as a partial replacement for mustard oil in the diet of healthy young Indians. Ind J Physiol Pharmacol. 49(1):49–56.

Sharma J P. Academic Biology IX. Laxmi Publications. p. 97. ISBN 978-81-7008-579-9. Retrieved, 2012.

Sharma S, De Roose E, Cao X et al. 2009. Dietary intake in a population undergoing a rapid transition in diet and lifestyle: the Inuvialuit in the Northwest Territories of Arctic Canada. Can J Pub Health. 100: 442–448.

Sheftel J, Gannon BM, Davis CR, Tanumihardjo SA. 2017. Provitamin A biofortified maize consumption increases serum xanthophylls and 13C-natural abundance of retinol in Zambian children. Exp Biol Med. (Maywood). 242:1508–14.

Shekar M. 2015. Investing In Nutrition: The foundation for development and investment framework to reach the global nutrition targets. Washington, DC.

Sigh S. 2007. From exotic spice to modern drug?Cell130: 765–768.

Siri-Tarino P, Sun Q, Hu F, et al. 2010. Saturated fat, carbohydrate, and cardiovascular disease. The Amer J Clin Nutr. 91(3): 502-509.

Slattery M, Edwards S, Boucher K, Anderson K and Caan B. 1999. Lifestyle and colon cancer:an assessment of factors associated with risk. Am J Epidemiol. 150:869-77.

Smith BB, Nancy ECG, Erica McLeod, Karen M, Meghan N and Elizabeth. 2013. COMPASS: Navigating the rules of scientific engagement. PLoS biol. 11. e1001552. 10. 1371/journal. pbio. 1001552.

Sonibare M and Gbile Z. 2008. Ethnobotanical survey of anti-asthmatic plants in south western Nigeria. Afri J Tradl Compl Altern Med. 5, 340. 10. 4314/ajtcam. v5i4. 31288.

Sraboni Esha andMalapit, Hazel and Quisumbing, Agnes and Ahmed, Akhter. 2014. Women's Empowerment in Agriculture:What role for food security in Bangladesh? World Development. 61:11–52. 10. 1016/j. worlddev. 2014. 03. 025.

Srinivasan, K. 2014. Antioxidant potential of spices and their active constituents. Crit Rev Fd Sci Nutr. 54: 352–372.

Stamoulis K and Zezza A. 2003. ESA Working Paper No. 03-17 (online). Rome, FAO. http:// www. fao. org/3/ae050e/ae050e00. pdf).

Start D and Hovland I. 2004. Tools for Policy Impact:A Handbook for Researchers. London, Overseas Development Institute.

Statistics Botswana. 2015. Annual Agricultural Survey Report. 2013. Gaborone: Statistics Botswana. http://www. statsbots. org. bw/sites/default/files/publications/ agricreport2013. pdf

Stevens GA. et al. 2015. Trends and mortality effects of vitamin A deficiency in children in 138 low- income and middle- income countries between 1991 and 2013: a pooled analysis of population -based surveys. 3(9): PE 528-536

Storozhenko S, De Brouwer V, Volckaert M, Navarrete O, et al. 2007. Folate fortification of rice by metabolic engineering. Nat Biotechnol. 11:1277–9.

Strengthening sector policies for better food security and nutrition results: political economy analysis. Policy Guidance Note 8 by FAO and EU. 2017, Publisher: FAO, European Union; http://www. fao. org/3/a-i7212e. pdf

Subodh Kapoor. 2002. The Indian Encyclopaedia. Cosmo Publications. p. 1745. ISBN 978-81-7755-257-7. Retrieved, 2012.

Subramaniam M D, Vellingiri B S. 2014. In: Lee and In Ho Kim. Korean J Fd Sci Anim Resour. v. 34(4). PMC4662155.

Sugishima Yuki (National Registered Dietician/ Certified Sports Dietician), 2019. Assistant Professor, Shigakkan University, Faculty of Wellness, Department of Nutrition, Joint certified by The Japan Dietetic Association and The Japan Sport Association.

Susie J T and Ke L. 1993. Women Writing in India:The Twentieth century. Feminist Press. p. 422. ISBN 978-1-55861-029-3. Retrieved 28June 2012.

Szabó IL, Czimmer J, Szolcsányi J, Mózsik G. Curr Pharm Des. 2013. 19(1):84-9. Molecular pharmacological approaches to effects of capsaicinoids and of classical antisecretory drugs on gastric basal acid secretion and on indomethacin-induced gastric mucosal damage in human healthy subjects (mini review).

Tacon A. 1997. Fishmeal replacers: Review of antinutrients within oilseeds and pulses-A limiting factor for the aquafeed Green revolution?Feeding Tomorrow's Fish. 22.

Takahashi C, Kikuchi N, Katou N, MikiT, YanagidaF and Umeda M. 1995. Possible anti-tumour-promoting activity of components in Japanese soybean fermented food, Natto: effect on gap junctional intercellular communication. Carcinogenesis. 16: 471–476.

Talsma E F et al. 2013. Biofortified cassava with pro-vitamin a is sensory and culturally acceptable for consumption by primary school children in Kenya. Plos One. https://doi. org/10. 1371/journal. pone. 0073433.

Tan Su-Lyn. 2018. Malaysia & Singapore. Lonely Planet. Retrieved 23 March 2018 – via Google Books.

Tang G, Gu X, Xu Q, Zhao X, et al. 1999. Green and yellow vegetables can maintain body stores of vitamin A in Chinese children. Am J Clin Nutr. 70:1069–76. http://ajcn. nutrition. org/ content/70/6/1069.

Tang G, Hu Y, Yin S, Wang Y, Dallal G, Grusak M, Russell R. 2015. Beta-carotene in Golden Rice is as good as beta-carotene in oil at providing vitamin A to children (retraction of vol 96, pg 658, 2012). Am J Clin Nutr. ;102(3):715. doi:10. 3945/ajcn. 114. 093229.

Tang G, Qin J, Dolnikowski G, et al. 2009. Golden Rice is an effective Source of vitamin A. Am J Clin Nutr. 89:1–8. http://ajcn. nutrition. org/content/89/6/1776. short.

Tang G. 2013. Using plant foods rich in β-carotene to combat vitamin A deficiency. Sight Life. 27(1). http://www. sightandlife. org/fileadmin/data/Magazine/2013/27_1_2013/using_ plant_foods_ rich_in_beta_carotene_to_combat_VAD. pdf.

Tanumihardjo SA, Ball AM, Kaliwile C and Pixley KV. 2017. The research and implementation continuum of biofortified sweet potato and maize in Africa. Ann N Y Acad Sci. 1390:88–103.

Tapsell LC, Hemphill I, Cobiac L, Patch CS, Sullivan DR, Fenech M, Roodenrys S, Keogh JB, Clifton PM, Williams PG, Fazio VA, Inge KE. 2006. Health benefits of herbs and spices: the past, the present, the future. Med J Aust. 185(4 Suppl):S4-24.

Tapsell L C. et al. 2006. Health benefits of herbs and spices: the past, the present, the future. The Med J Austr. 185: S4–S24.

Tarla Dalal. 2001. Achaar Aur Parathe. Sanjay and Co. p. 85. ISBN 978-81-86469-54-5. Retrieved , 2012.

Tarla Dalal. 2007. Punjabi Khana. Sanjay and Co. p. 6. ISBN 978-81-89491-54-3. Retrieved, 2012.

Taste of Mangalore. The Hindu. Chennai, India. 2002. Archived from the original 2012. Retrieved, 2012.

Tessema M, Gunaratna NS, Donato K, Cohen JL, Connell M, Belayneh D, et al. 2016. Translating the impact of quality protein maize into improved nutritional status for Ethiopian children:study protocol for a randomized controlled trial. BMC Nutr. 2:54.

TeufelAlessandraGalièNils, AmyWebbGirard, IsabelleBaltenweck, PaulaDominguez-Salas, Mindy, J. Price, RebeccaJones, BenLukuyu, LukeKorir, IlanaG. Raskind, KristieSmith and Kathryn M. Yount. 2019. Women's empowerment, food security and nutrition of pastoral communities in Tanzania. Global Food Security. 23: 125-134. https://doi. org/10. 1016/j. gfs. 2019. 04. 005

Thalassery Biriyani. 2012. mysingaporekitchen. com. Archived from the original on 21 September 2013. Retrieved, 2013.

The Sunday Tribune– Spectrum – Lead Article. 2003. The Tribune. Retrieved 2012.

The Telegraph. 2016. Panel seal on rasgulla's Odisha origin- Committee cites references from old texts to refute Bengal's claim. Archived from the original on 18 August 2017. Retrieved, 2017.

The United States (US) Dietary Guidelines. 2015-2020.

Thomas EL, Frost G, Taylor-Robinson SD, et al. 2012. Excess body fat in obese and normal-weight subjects. Nutr Res Rev. 25(1):150-61.

Tipilda A and Kristjanson P. 2009. Women and Livestock Development: A Review of the Literature. International Livestock Research Institute (ILRI) Innovation Works Discussion Paper 01–09. Nairobi: ILRI.

Tirado M C. 2011. Enhancing women's leadership to address the challenges of climate change on nutrition security and health. center for public health and climate change at the public health institute, World Food Programme (WFP), UN Standing Committee on Nutrition and Action Against Hunger (ACF). New York. United Nations.

TNN. 2009. Kodagu still swears by pork. The Times of India. Retrieved, 2012.

Tome D and Bos C. 2007. Lysine requirement through the human life cycle. J Nutr. 137:1642S–1645S.

Trichopoulou A, Soukara S and Vasilopoulou E. 2007. Traditional foods:a science and society perspective. Trends Fd Sci Technol. 18:420-427.

Trichopoulou A, Vasilopoulou E, Georga K, Soukara S and Dilis V. 2006. Traditional foods: Why and how to sustain them. Trends Fd Sci Technol. 17:498-504.

Trijatmiko K, Dueñas C, Tsakirpaloglou N, Torrizo L, et al. 2016. Biofortified Indica rice attains iron and zinc nutrition dietary targets in the field. Scientific reports, 6, article no. 19792. https://www. nature. com/articles/srep19792.

Tsen CC. 1976. Regular and protein fortified biscuits from composite flours. Cereals Fds World 21: 633- 634.

Tumwet T N, Kangethe E K, Kogi-Makau W and Mwangi A M. 2014. Diversity and immune boosting claims of some african indigenous leafy vegetables in western Kenya. Afri J Fd Agric Nutr Develop. 14 (1): 1684-5374

Udoh F V. 1999. Uterine muscle reactivity to repeated administration and phytochemistry of the leaf and seed extracts of Piper guineense. Phytotherapy Research: PTR13:55–58.

Udoh F V, Ekanem A P and Eyo V O. 2012. Pharmacodynamic effect of methanol extract of Piper guineense leaf on uterine physiology. Pharmacologia. 3: 200–203.

Udourioh G A and Etokudoh M F. 2014. Essential oils and fatty acids composition of dry fruits of Tetrapleura tetraptera. J Appl Sci Environl Manage. 18: 419–424.

Udry C, Alderman H, Hoddinott J and Haddad L. 2005. Gender differentials in farm productivity: implications for household efficiency and agricultural policy. Fd Pol. 20(5):407–423.

Udry C. 1996. Gender, agricultural production, and the theory of the household. J Pol Econ. 104(5):1010–1046.

Uhegbu FO, Imo C and Ugbogu A E. 2015. Effect of aquous extract of Piper guineense seeds on some liver enzymes, antioxidant enzymes and some hematological parameters in Albino rats. Internl J Pl Sci. 4: 167–171.

UN Department of Public Information (UNDPI). 2010. Fact Sheet on Millennium Development Goal 3. DPI/2650 C.

UN Economic and Social Council (ECOSOC). 2007. Strengthening efforts to eradicate poverty and hunger, including through the global partnership for development. Para. 14; World Food Programme (WFP) 2009a, p. 6. Report of the SecretaryGeneral. UN doc. E/2007/71. New York: ECOSOC.

UN Educational Scientific and Cultural Organisation (UNESCO). 1998. Bridging the Gap between Intention and Action: Girls' and Women's Education in South-Asia. Bangkok: UNESCO Regional Office for Asia and the Pacific. 2007.

UN SCN. 2004. Nutrition Information in Crisis Situations – Report Number I, 2004. United Nations System Standing Committee on Nutrition.

UN. 2010. Levels and trends in child mortality: estimates developed by the United Nations inter-agency group for child mortality estimation. http://www. childmortality. org/files_v20/download/Levels%20and%20Trends%20in%20Child%20Mortality%20Report%20 2010. pdf.

UN. 2015. Levels and trends in child mortality. http://www. who. int/maternal_child_adolescent/documents/levels_trends_child_mortality_2015/en/.

UN. 2015. UNAIDS fact sheet 2015. http://www. unaids. org/en/reSources/campaigns/HowAIDSchangedeverything/factsheet.

UN. Millennium development goals. 1990–2015. undated. http://www. unfoundation. org/what-we-do/issues/mdgs. html?referrer=https://. Accessed 25 Sept 2017.

UNICEF, 2007. A World Fit for us: The Children's Statement from the UN Special Session on Children: Five years on.

UNICEF, WHO, The World Bank. 2020. UNICEF/WHO/WB Group. Global Database on Child Growth and Malnutrition.

UNICEF. 1990. World declaration on the survival, protection and development of children. http://www. unicef. org/wsc/declare. htm.

UNICEF. 2004. The micronutrient initiative. Vitamin and mineral deficiency. A global damage assessment report. https://www. unicef. org/media/files/davos_micronutrient. pdf.

United Nations General Assembly. 2000. United Nations millennium declaration. A/RES/55/2. UNGeneral Assembly. https://www. unescap. org/reSources/ga-resolution-552-united-nations-millennium-declaration.

United Nations, New York, 2008. Department of Economic and Social Affairs Office for ECOSOC Support and Coordination Achieving Sustainable Development and Promoting Development Cooperation Dialogues at the Economic and Social Council.

United Nations. 2015. The millennium development goals report. 2015. http://www. un. org/millenniumgoals/2015_MDG_Report/pdf/MDG%202015%20rev%20(July%201). pdf.

Uusiku N P, A Oelofse A, Duodu K G, Bester M J and Faber M. 2010. Nutritional value of leafy vegetables of sub-Saharan Africa and their potential contribution to human health: A Review. J Fd Comp Anal. 23: 499-509.

Uyom EA, Ita E E and Nwofia G E. 2013. Evaluation of the chemical composition of Tetrapleura tetraptera (Schum and Thonn.)Taub. Accessions from river state Nigeria. Internl J Med Arom Pl. 3:386–394.

Uzoejinwa B and Ani, Ozoemena and Abada, Ugochukwu and Ugwuishiwu, Boniface and Jude and Ohagwu and Nwakaire J. 2016. Small-scale food processing enterprises: measures for national development and addressing food security challenges in Nigeria. Internal J Sci Tech Res Eng. 1. 72-82.

van Jaarsveld PJ, Faber M, Tanumihardjo SA, Nestel P, Lombard CJ, Benade AJ, et al. 2005. Beta-carotene-rich orange-fleshed sweet potato improves the Vitamin A status of primary school children assessed with the modified-relative-dose-response test. Am J Clin Nutr. 81:1080–7.

Vanhonacker F, Verbeke W, Lengard V et al. 2008. Consumer-based definition and general image of traditional foods in Europe. In: Perspectives of Traditional Food Supply Chains on the European Market. Proceedings of 12th Congress of the European Association of Agricultural Economists, 'People, Food and Environments: Global Trends and European Strategies'. Ghent, Belgium.

Vanhonacker Filiep and Kühne, Bianka and Gellynck, Xavier and Guerrero, Luis and Hersleth, Margrethe and Verbeke, W. 2013. Innovations in traditional foods: Impact on perceived traditional character and consumer acceptance. Fd Res Internl. 54. 10. 1016/j. foodres. 2013. 10. 027.

Verbeke, W. 2005. Agriculture and the food industry in the information age. European Review of Agriculture Economics. 32(3), 347-368.

Viles S, Nikam ST and Yemul VL. 1995. Oral hypoglycaemic agents from indigenous medicinal plants. Proc of World Congression. Biotechnological developments in medicinal substances of plant and marine origin.

von Braun J. 2008. Rising Food Prices: What should be done? Eurochoices 7(2):30–35.

Vorster H J and Jansen van Rensburg W S. 2005. Traditional Vegetables as a Source of Food in South Africa. Afri Crop Sci Conf Proceedings 7: 669-72

Wactawski-Wende J, Kotchen J M, Anderson GL, Assaf AR, Brunner RL, O'Sullivan MJ, et al. 2006. Calcium plus vitamin D supplementation and the risk of colorectal cancer. N Engl J Med. 354:684-96.

Welch RM, Graham RD. 2004. Breeding for micronutrients in staple food crops from a human nutrition perspective. J Exp Bot. 55:353–64.

West CE. 2000. Vitamin A and measles. Nutr Rev. 58:S46–54.

West K, Klemm R, Sommer A. 2010. Sound science, sound policy. J World Public Health Nutr Assoc. 1(5):211–229. http://www. wphna. org/htdocs/downloadsoctober2010/10-10%20 WN%20Comm%20VitA%20west%20et%20al. pdf.

Whitcher J, Srinivasan M and Upadhyay M. 2001. Corneal blindness: a global perspective. BullWorldHealthOrgan. 79(3):214-221. http://www. who. int/bulletin/archives/79(3)214. pdf.

White Paper on International Development:Eliminating world poverty: Making governance work for the poor. 2006. UNGEI. United Nations girls education initiative.

WHO (World Health Organization). 2009. Global Health Risks. Mortality and burden of disease attributable to selected major risks.

WHO, Ranjit R C and Uton M R. 2001. Traditional medicine in Asia. SEARO Regional Publications No. 39 © World Health Organization. ISBN 92 9022 2247.

WHO. 2007. Joint WHO/FAO/UNU Expert Consultation.

WHO. 2008. The Commission on Social Determinants of Health - what, why and how?

WHO. 2011. Guideline:neonatal vitamin A supplementation. 1–25. http://apps. who. int/ iris/bitstream/10665/44626/1/9789241501798_eng. pdf. WHO. Guideline: vitamin A supplementation in infants and children 6–59 months of age. 2011. p1–25. http://apps. who. int/iris/bitstream/10665/44664/1/9789241501767_eng. pdf.

WHO. 2013. Nutrition, micronutrient deficiencies, vitaminA deficiency. http://www. who. int/ nutrition/topics/vad/en/.

Why are banana leaf meals popular in Malaysia. asianinspirations. com. au. Retrieved, 2018.

Wiggins S, Farrington J, Henley G, Grist N and Locke A. 2013. Agricultural development policy: a contemporary agenda. Background Paper for GIZ. Young E. and Quinn L. 2012. Making Research Evidence Matter. A guide to policy advocacy in transition countries. Open Society Foundations, Budapest.

Williamson C. 2006. The different uses of food composition databases. Synthesis Report No 2, British Nutrition Foundation, European Food Information ReSource Consortium, Norwich.

Winzenberg T and Jones G. 2013. Vitamin D and bone health in childhood and adolescence. Calcified Tissue Internl (Review). 92 (2): 140–50.
Wolf WJ and Cowan JC. 1975. Soybeans as a Food Source. CRC Press, Inc, Cleveland.
World Cancer Research Fund (WCRF) / American Institute for Cancer Research (AICR). 2007. Food, nutrition, physical activity and the prevention of cancer: a global perspective. Washington DC. USA.
World Health Organization (WHO) Media Center (2013). Fact sheet N°317: Cardiovascular diseases (CVDs).
Wu J, Micha R, Imamura F, et al. 2012. Omega-3 fatty acids and incident type 2 diabetes: a systematic review and meta-analysis. The Brit J Nutr. 107(Suppl 2): S214-S227.
Wu K, Willett WC, Fuchs CS, Colditz GA and Giovannucci EL. 2002. Calcium intake and risk of colon cancer in women and men. J Natl Cancer Inst. 94:437-46
www. casemine. com
www. eurofir. net
www. fao. org/contact-us/licence-request
www. fao. org/publications
www. fda. gov/food/ingredientspackaginglabeling/gras/scogs/ucm261245. htm
www. harvest plus. org/content/food-crisis
www. holidify. com.
www. rheon. com/index. php
www. wilsonsdisease. org
Yadava DK, Choudhury PR, Hossain F Kumar D. 2017. Biofortified varieties: Sustainable way to alleviate malnutrition. New Delhi: Indian Council of Agricultural Research.
Yadava DK, Hossain F and Mohapatra T. 2018b. Nutritional security through crop biofortification in India: Status & future prospects. Indian J Med Res. 148(5): 621–631.
Yadava DK, Hossain F, Choudhury PR, Kumar D, Singh AK. 2018a. Biofortification of crops: A sustainable tool for nutritional security in India. Indian Farming. 68:37–42.
Ye X, et al. 2000. Engineering the provitamin A β-carotene biosynthetic pathway into (carotenoid-free) rice endosperm. Science. 287(5451):303–5. https://www. ncbi. nlm. nih. gov/pubmed/10634784, http://www. goldenrice. org/PDFs/Ye_et_al_Science_2000. pdf.
Yeager D S and Walton G M. 2011. Social-psychological interventions in education: They're not magic. Rev Educnl Res. 81(2): 267–301.
Yimer F and Tadesse F. 2015. Women's Empowerment in Agriculture and Dietary Diversity in Ethiopia. Ethiopia Strategy Support Program (No. 80).
Young and FAO. 2015. Enhancing the nutritional impact of agriculture investment programmes:A checklist and guidance for programme formulation (pages 6-12).
Young. 2013. Inequality, the urban-rural gap, and migration. The Quart J Econ. 128(4):1727–1785.
Yvette B. 2018. Proteins and health. Medically reviewed by Katherine Marengo, LDN, RD, specialty in nutrition.
Zhou Y, Tian C and Jia C. 2012. Association of fish and n-3 fatty acid intake with the risk of type 2 diabetes: a meta-analysis of prospective studies. The Brit J Nutr. 108(3): 408-417.
Ziemann M. 1999. Internationalisierung der Ernährungsgewohnheiten in ausgewählten europäischen Ländern. In: Europäische Hochschulschriften, Reihe V, Volks- und Betriebswirtschaft, Bd. 2470. Frankfurt/M.
Zuma MK, Kolanisi U and Modi AT. 2018. The potential of integrating provitamin A-biofortified maize in smallholder farming systems to reduce malnourishment in South Africa. Int J Environ Res Public Health. 15 pii: E805 .